AF770686

Principles of Agronomy and Agricultural Meteorology

Principles of Agronomy and Agricultural Meteorology

Edited by:

Kailash Rameshrao Malode

Delve
PUBLISHING

www.delvepublishing.com

Principles of Agronomy and Agricultural Meteorology

Kailash Rameshrao Malode

Delve Publishing
224 Shoreacres Road
Burlington, ON L7L 2H2
Canada
www.delvepublishing.com
Email: orders@arclereducation.com

© 2021 Delve Publishing

ISBN: 978-1-77407-736-8 (Hardcover)

This book contains information obtained from highly regarded resources. Reprinted material sources are indicated and copyright remains with the original owners. Copyright for images and other graphics remains with the original owners as indicated. A Wide variety of references are listed. Reasonable efforts have been made to publish reliable data. Authors or Editors or Publishers are not responsible for the accuracy of the information in the published chapters or consequences of their use. The publisher assumes no responsibility for any damage or grievance to the persons or property arising out of the use of any materials, instructions, methods or thoughts in the book. The authors or editors and the publisher have attempted to trace the copyright holders of all material reproduced in this publication and apologize to copyright holders if permission has not been obtained. If any copyright holder has not been acknowledged, please write to us so we may rectify.

Notice: Registered trademark of products or corporate names are used only for explanation and identification without intent of infringement.

Delve Publishing publishes wide variety of books and eBooks. For more information about Delve Publishing and its products, visit our website at www.delvepublishing.com.

ABOUT THE EDITOR

Dr. Kailash Malode (1980) is presently serving as Assistant Professor, Department of Soil Science and Agriculture Chemistry, college of Agriculture, Parul University, Vadodara, Gujarat. He obtained his B. Sc.(Ag.) in 2006 from Marathwada Agriculture University, Parbhani. M. Sc. (Ag.)2009 in Dr. Panjabrao Deshmukh Krishi Vidyapeeth,Akola and Ph. D. Soil Science And Agriculture Chemistry in 2014 from Vasantrao Naik Marathwada Krishi Vidhyapeeth, Parbhani. He started his career as Assistant professor in 2014. His field of specialization is soil fertility. He has published 09 research paper in National and international journals of reputed. He has participated many National, State Seminar and symposiums. He has written 02 practical manuals for under graduate course for the benefit of students besides several Radio talk, Lectures in farmer traning programme. Dr. Kailash has also written 16 popular articles in Shati Bharti, Shati pragati and other reputed Magazines. He has Best writer award on 2015 for Krishi dooth popular Magazines of Maharashtra

TABLE OF CONTENTS

GLOSSARY

A

Aberrations – A departure from what is normal, usual, or expected, typically an unwelcome one.

Acreage – An area of land and its buildings, used for growing crops and rearing animals.

Ancillary – Providing necessary support to the primary activities or operation of an organization, system, etc.

Aphids – A small bug which feeds by sucking sap from plants; a black fly or greenfly.

B

Blotters – A sheet or pad of blotting paper in a frame.

C

Commodities – A raw material or primary agricultural product that can be bought and sold, such as copper or coffee.

Conspicuous – Attracting notice or attention.

Contour – An outline representing or bounding the shape or form of something.

D

Discord – Disagreement between people.

Disparate – Composed of inherently different or distinct elements.

Distrust – The feeling that someone or something cannot be relied upon.

Domesticated – Tame and kept as a pet or on a farm.

Driveway – A short road leading from a public road to a house or other building.

E

Elasticity – Elasticity is the measurement of the proportional change of an economic variable.

Elevate – Raise or lift (something) to a higher position.

Equilibrium – A state in which opposing forces or influences are balanced.

Evapotranspiration – The sum of evaporation and plant transpiration from the Earth's land and ocean surface to the atmosphere.

F

Fallow – plowed and harrowed but left for a period without being sown to restore its fertility or to avoid surplus production.

Fertilizer – A natural substance that is used to make the ground more suitable for growing plants.

Fodder – Animal feed.

Forage – To feed on grass or herbs.

Foster – Encourage the development of (something, especially something desirable).

Furrows – To develop creases or folds.

G

Germinate – To grow or cause to grow.

Gigantic – Of very great size or extent.

H

Hopper – A plant-eating insect with long hind legs which are used for jumping and for producing a chirping sound, frequenting grassy places and low vegetation.

I

Imbibition – The absorption of one substance by another, in particular the uptake of water by a plant or seed.

Infestation – The presence of an unusually large number of insects or animals in a place, typically so as to cause damage or disease.

Influenced – The capacity to have an effect on the character, development, or behaviour of someone or something, or the effect itself Invasion.

L

Laborious – Requiring considerable time and effort.

M

Machetes – A broad, heavy knife used as an implement or weapon, originating in Central America and the Caribbean.

Manifest – Clear or obvious to the eye or mind.

Manipulate – Handle or control (a tool, mechanism, information, etc.) in a skillful manner.

N

Nomadic – Living the life of a nomad; wandering.

P

Paired Row Planting – Two cane rows are brought together followed by a wide gap before the next set of two rows.

Panacea – A solution or remedy for all difficulties or diseases.

Peat – An accumulation of partially decayed vegetation or organic matter.

Plant geometry – Arrangement of the plants in different rows and columns in an area to efficiently utilize the natural resources.

Precipitation – Any product of the condensation of atmospheric water vapor that falls under gravity from clouds.

Prosperity – Successful in material terms; flourishing financially.

R

Random plant geometry – No equal space is maintained in plantation.

Rearing – Bring up and care for a plant until they are fully grown.

Rectangular method of sowing – Trees are planted on each corner of a rectangle.

Remunerative – Financially rewarding; lucrative.

Revolution – A forcible overthrow of a government or social order, in favor of a new system.

S

Seed box – Also called trays or flats, offer gardeners the opportunity to start a large number of seeds in one container.

Seed drill – a device that sows the seeds for crops by positioning them in the soil and burying them to a specific depth.

Seedbed – A place or environment that favors the development of something.

Sequestration – The state of being alone or kept apart from others.

Shovel – A tool for digging, lifting, and moving bulk materials, such as soil, coal.

Skip row planting – One or more rows are omitted in between planted rows.

Sod – The surface of the ground, with the grass growing on it; turf.

Sorghum – A cereal grain that grows tall like corn, and it is used for a lot more than just sweetening.

Sparsely – In a thinly dispersed manner; in small numbers.

Square plant geometry – Plants are sown at equal distances on either side.

Stubble – The cut stalks of cereal plants left sticking out of the ground after the grain is harvested.

Subsistence – The action or fact of maintaining or supporting oneself, especially at a minimal level.

T

Tiller – A horizontal bar fitted to the head of a boat's rudder post and used for steering.

Treble – Consisting of three parts; threefold.

V

Viability – Ability to work successfully.

W

Weather – The state of the atmosphere at a particular place and time.

Weather forecast – An analysis of the state of the weather in an area with an assessment of likely developments.

LIST OF FIGURES

LIST OF TABLES

LIST OF ABBREVIATIONS

ASRs	Agriculturally Significant Rains
BST	Bovine Somatotropin
CRM	Climate Risk Management
DP	Dependable Precipitation
DSSs	Decision Support Systems
EPA	evaporative power of air
EPIC	Economics and Policy Innovations for Climate
ET	evapotranspiration
Ex-Act	ex-ante carbon balance tool
FDA	Food and Drug Administration
GISs	geographic information systems
GLEAM	global livestock environmental assessment model
GMOs	genetically modified organisms
HSU	heat stress unit
INM	integrated nutrient management
IPCC	Intergovernmental Panel on Climate Change
IPM	integrated pest management
LEISA	low external input sustainable
LRF	long range forecast
MAGHG	monitoring and assessment of greenhouse gas emissions
MCC	mesoscale convective complexes
MOSAICC	Modeling System for Agricultural Impacts of Climate Change
MRF	medium range forecast
NC	now-casting
NMHSs	National Meteorological and Hydrological Services
NWP	numerical weather prediction models
P	precipitation

PET	potential evapotranspiration
RH	relative humidity
SAFA	Sustainability Assessment of Food and Agriculture
SIT	sterile insect technique
SO2	sulfur dioxide
SRF	short-range forecast
UNFCCC	United Nations Framework Convention on Climate Change
VSRF	very short-range forecast
WTO	World Trade Organization
WMO	World Meteorological Organization

PREFACE

Agriculture sector is a huge contributor to the world economy and is an integral part of the sources of the food and nourishment for the people. It is the first answer for the basic needs of the people and thus, its importance is undoubtedly a lot.

The field of agriculture involves a lot of study and operations as there is a constant push to develop this field and get the best output from the sector, which must be enough to satisfy the nourishment and nutritional demands of the people and capable to support the whole population, avoiding any possibility of hunger or malnutrition.

Agriculture attracts interest from scholars and researchers across several fields. There are many individuals who are working in the direction of uplifting the agriculture sector and a major contribution is from the economists as agriculture has a major impact on the economic conditions of the whole world.

This book tries to make the readers aware about the various aspects related to agronomy. It educates the readers about the terminologies that relate to the subject of agronomy and defines other branches of agriculture in brief.

The readers are informed about the major crops that control the agriculture sector, the types of farming and the agricultural practices in different regions. It elaborates exclusively on the subject of agronomy, defining its scope and informing about the definitions of agronomy given by the scholars.

The book further elaborates on the subject of agricultural meteorology making the readers informed about the history of meteorology, the benefits of meteorology in agriculture sector and defining the various meteorological parameters in the field of agriculture.

Further, in Chapter 4 the readers are informed about the concepts of radiation, temperature, and climate and relate them to agricultural practices. In the next chapter, the book educates the readers about the manner in which the cropping is done and talks about the cropping systems that are adopted by the farmers and organizations across the world.

Then the readers are told about the importance of geometry in the planting of the crops and the manner in which the geometry affects the growth of various crops. They are also taken through the manner in which weather affects the growth of crops and how the factors pertaining to weather conditions can be managed to have profitable yield. The book also highlights the importance of weather forecasts and lists ways in which the farmers can use these forecasts to their advantage.

In the concluding chapter, the book emphasizes on the manner in which climate change is impacting agriculture and suggests ways to control climate change as well as mitigate the effects of such changes taking place around the globe.

I hope that this book proves informative in the way that the readers can refer to this book for any piece of knowledge regarding agronomy and the meteorology in agriculture sector. I also hope that this book can be used to educate the subject enthusiasts about the field of agronomy and agricultural meteorology in depth.

CHAPTER 1

Introduction to Agriculture and Its System

CONTENTS

This chapter is about the importance of agriculture in our society, their various types, their importance and their systems or methods that are used in different regions according to the environment condition.

This chapter explains all the branches of agriculture that are currently present, like crop production, horticulture, forestry, agriculture engineering, home science, fishery science, etc.

The farming types are also explained in this chapter with the major crops that are grown worldwide and are commonly used by the people.

In this chapter, it is also explained the different agriculture systems that are present or adopted by different regions according to the environmental conditions and requirements of the crop that is to be grown. In this part, the areas that are using these systems and the characteristics of these systems are explained.

1.1. INTRODUCTION

It has been over 10,000 years since the agriculture practice began to develop. There are many people around the world who have discovered the value of food, either it is wild plants or animals. They also developed methods to domesticate them and bred them.

The most important crops that are being used nowadays are wheat, rice, corn, barley, rye, sugarcane, and sugar beet. Whereas in case of animal food the most used are goat, sheep, pigs, cattle, poultry which includes chickens, ducks, turkey, etc.

They also used animal products such as milk, cheese, eggs, oils, and nuts. There are some major foods that are being used by humans, include fruits, vegetables, and olives. There are also feed grains for the animals, this basically includes field corn, sorghum, and soybeans.

Agriculture income is being derived from nonfood crops. These crops include rubber, tobacco, oil seeds, and fiber plants. These crops are basically used as synthetic chemical compounds. There are few animals that are basically raised for the pelts that are animal skins.

Climate, water supply, terrain, and ecology are some of the factors or conditions that help in determining what is raised in an area. In the year 2003, there was about 44% of the world's labor force was employed in agriculture practices.

The distribution of land ranges from 66% to where economically active population in sub-Saharan Africa. Whereas, in the United States and Canada it is less than 3%. It is about 60% in the case of Asia and the Pacific, 19% is in Latin America and the Caribbean, and about 9% in Europe.

There are also Farm sizes that vary basically from region to region. In the early 2000s, the average Canadian farm was about 273 hectares which is approximately 674 acres per farm. In the case of the United States, the area of field is about 180 hectares which approximately 440 acres. In case of India, the size of the single landholding is of 2 hectares which is about 5 acres.

The size of agriculture land also depends on the purpose of the farm. In the case of large holding, commercial farming or production for cash is said to usually take place.

Latifundia which is of Latin America is a large, privately-owned estate those are worked by the tenant labor. Single crop plantation production involves tea, cocoa, and rubber. Wheat farming is considered as the most efficient one when they comprise thousands of hectares. The teams of people and machines can work the wheat farms.

Australian sheep stations and other livestock farms are basically kept large so as to provide grazing for the thousands of the animals. Before the practice of individual subsistence farming or small family mixed farming, the operation was in large numbers.

But now, these practices or operations are decreasing in the number in developed countries. Whereas there are still some developing countries that are still practicing these farming operations, these countries are Africa and Asia.

Nomadic headers range over a large area in a sub-Saharan Africa, Afghanistan, and Lapland. Lapland is the region that is largely present within the arctic circle which basically extends across the northern parts of Norway's, Finland, Sweden, and in the Kola Peninsula of Russia.

In the area like Mongolia, herding is considered as a major part of agriculture. There are many countries whose foreign exchange is being earned by simply deriving from a single agricultural commodity. These countries are, Sri Lanka which totally depends on tea, Denmark is specialized in a dairy product, New Zealand and Argentina for meat production, Australia for wool production.

But there some other countries whose foreign exchange earns depends on many crops like in the case of the United States which is producing wheat, corn, soybean as a major commodity in recent decades and is responsible for the foreign exchange earnings.

Agriculture provides various materials, food, feed, and fuel, and furniture, the raw material for and from the factories. Agriculture also provides a free fare and fresh environment, favor for the friendships by simply eliminating the fights; it also provides abundant food for driving out the famine.

It is considered that a satisfactory agricultural production will bring peace, prosperity, health, harmony, and wealth in the life of the individual of the nation. These all satisfactory is being bought by simply driving away from the distrust, revolution, and discord.

It also helps in elevating the community by simply consisting of different castes and clauses. Therefore, this leads to better social, political, economic, and cultural life. Agriculture development is considered as a multidirectional which is having a galloping of speed and there is a rapid spread with respect of time and space.

After the green revolution takes place, farmers started using improved cultural practices and agricultural inputs. They used in the intensive cropping system with the laborer intensive programs to enhance the production potential per unit land, input, and time.

It also provides a suitable environment for all these improved genotypes. This is done to foster and manifest their yield potential in newer areas and seasons. The agriculture consists of growing plants and rearing animals. These agriculture practices help to yield, produce, and help to maintain the biological equilibrium in nature.

There are over 10,000 years since agriculture began to be developed. The peoples everywhere have discovered the food value of wild plants and animals, and they also domesticated them and breed them.

The most important crops are cereals such as wheat, rice, barley, corn, and rye; sugarcane and sugar beets; meat animals such as sheep, cattle, goats, and pigs or swine; poultry such as chickens, ducks, and turkeys; animal products such as milk, cheese, and eggs; and nuts and oils. Fruits, vegetables, and olives are also major foods for people.

Feed grain that is used for the animals includes sorghum, soybeans, and field corn. Non-food crops such as fiber plants, rubber, tobacco, and oil seeds are used in synthetic chemical compounds and they also being used animal raising for the pelts i.e., for animal skins. From these non-food crops the agriculture income is also being derived.

There are several conditions that can determine what is to be raised in an area which includes animals, water supply, terrain, ecology, and water works. In the year 2003, there are about 44% of the world's labor force that is being employed in the agriculture field.

There is about 66% of the economically active population in the distribution range in the areas of sub-Saharan Africa especially in the area of Mali, Ethiopia, Zimbabwe, etc. Whereas, in the United States and Canada, the active population is less than 3%. It is about 60% of the figure is present in the areas of Asia and the pacific. In the areas of Latin America and the Caribbean, this figure ranges to 19% and in Europe, this figure is about 9%.

The farm size widely varies from region to region. The farm ranges about 273 hectare that is about 675 acres per farm in the early 2000s for the average Canadian farms, whereas, for the United States the farm size is about 180 hectares, that is about 440 acres. When it comes for India, the average size of a single landholding is about 2 hectares which is about 5 acres.

The size of the farm also depends on the purpose for which the farm land is used. Commercial farming, or production for cash, usually takes place on large holdings. The latifundia of Latin America are large and are privately-owned estates worked by tenant labor. Single-crop plantations produce tea, rubber, and cocoa.

Wheat farms can be worked by teams of people and machines which are the most efficient when they comprise thousands of hectares. Australian sheep stations and other livestock farms must be large to provide grazing for thousands of animals.

Individual subsistence (condition of managing to stay alive) farms or small-family mixed-farm operations are decreasing in number in developed countries but are still numerous in the developing countries of Africa and Asia.

The Nomadic herders are said to range over the large area of lapland whose region mostly comes within the arctic circle that extending to the Northern parts of Norway, Finland, and Sweden, Afghanistan, and Africa. Areas such as Mongolia, herder is considered as a major part of agriculture.

1.2. HISTORY OF AGRICULTURE

As it is already known that, life began to develop around 3.5 billion years ago in the solar system. But the life forms on the planet around 4.6 billion years ago in the universe. Since then, evolution has produced about hundreds of millions of living species. In these species, many of the species are those that have disappeared in a course of time.

The first species that appear were plants. From which there are still more than 500,000 species that are still alive today. The plants, animal species are the one that are nearly millions in number. Not all the living species that have been identified and it is believed that every year a new species are being discovered.

The species that are living in a particular place their totality at a given moment in time form a population of the particular species. The totality of the plants and animals' populations is said to form a biocenosis for them who live in that particular place. The biocenosis and the inanimate environment or biotope weather it is geologically, morphologically or climatic that inhabits and forms an ecosystem.

It is being considered that all the ecosystem of the plant forms the ecosphere. It is also being considered that all the living beings, either it's animal or plants are being formed from the organic matter, water, and other minerals.

These organic matters are being formed from complex molecules like sugar, proteins, fats, and nucleic acid. These organic matters besides forming living beings are also used as the source of the energy which is necessary for life and it is also important for reproduction.

The plants are the autotrophs which simply means that they are capable of synthesizing their organic substance from the water, carbonic, and other elements by the means of the solar system. These organic substances are found in the atmosphere and the soil. While in the case of animals and humans, they are believed that they do not have this ability as they are heterotrophs.

Humans and animals are considered that they live upon the organic matter that is being provided directly by the animals. It is believed that they have been first consumed and are be assimilated. The biomass of an ecosystem is defined as the total mass of the organic matter that it contains. The organic matter includes waste product and excrement.

It is being considered that only plant species are productive of biomass. Whereas, humans and animals are believed that they do not produce the biomass. They can only feed upon it and can transform it. They are considered as an exploitive species.

This is the reason why the fertility of an ecosystem which includes its capacity to produce the biomass, is measured by the capacity to produce the plant biomass. Most of the animals are considered as a simple predator.

These animals are the content with obtaining their food by force. This force is done on the wild species of the plants and animals which are being exploited.

Some of the species among them are considered to provide a service for the exploited species. For example, bee transports the pollen which it usually gathers from the flower. After that they facilitate it's for the fertilization process. Millions of years before, the evolution produced millions of species of ants and termites. These species then further cultivated fungi or they even raised aphids.

These fungi and aphids are the domestic ones that are being exploited intensely by the ants and the termites through constant work. These constant works include managing the environment by multiplying the population and they also promote their development.

Humans are being considered as much more recent species and not as a born to be farmers or stock breeders. They are different from the ants and termites. They become like this after hundreds of millions of years of hominization. These years include biological, technical, and Cultural Revolution. It was less than 10,000 years ago when humans began to cultivate the plants and also started to breed the animals that they had domesticated. This happened in the Neolithic era.

With all this, they also introduced these plants and animals into all sorts of environments. In that environment, they are supposed to attempt to propagate them. In this way, the original natural ecosystem was transformed into cultivated ecosystem. They are also being fabricated and are being exploited by human care and by the attention.

From then on, human agriculture has been taken up the whole world. Human agriculture has become the principal factor in the transformation of the ecosphere. It also gains in the production and productivity that have been respectively influenced by the increase in the number of people and by the development of the social groups.

The main aim is to respond to three essential questions:—At what moment in the process of hominization did humans become farmers and why?—What is agriculture as a particular relation between living species?—from then, what is the role played by agriculture in the historical development of humanity?

These social groups are believed that they cannot produce their food. The main intention is to situate agriculture in the evolution of the life and the history of the humanity.

1.3. DEFINING AGRICULTURE

The word agriculture is being derived from the Latin word-Ager or Agri. which means soil and culture which directly gives the meaning cultivation of soil (Figure 1.1).

Figure 1.1. Image representing agriculture field.

Source: Image by Needpix.

One can consider that farming is a complete system which includes input, processing, and outputs. Where the input is the seeds, fertilizers, machinery, which undergoes operations.

In the modern terms, agriculture word comprises "the art and science of cultivating the soil, growing crops and rearing livestock."

These operations include plowing, sowing, irrigation, harvesting, and weeding. After the input, the desired output is received as a final product which includes crops, poultry products, dairy, etc.

1.4. TERMINOLOGY IN AGRICULTURE

As agriculture, the term is being derived from the words Ager and Culture. Where Ager means land or field and culture defines cultivation. Therefore, the term agriculture means the cultivation of land. Which simply means that it is an arch and science of producing crops and livestock for the economic purpose.

It can be also referred to as the science that involves the production of crops and livestock, which is derived from the natural resources of the earth. The primary aim of agriculture is to prepare the land to achieve more crops and at the same time, it is also kept in mind to protect the land from deteriorating and misuse. It is considered as synonymous with farming that includes the production of food, fodder, and other industrial materials.

The definition of agriculture was defined in the Agriculture Act of 1947. This act includes horticulture, growth of food, and growth of a seed, dairy farming, and livestock breeding and keeping. With this, they are also using land as grazing land, meadows land, marketing gardens, osier land, and nursery ground. The uses of land for the woodlands is where that use ancillary to the farming land for the Agriculture purpose.

It can be also defined as the purposeful work through which the elements in nature are being harnessed. This process is done to produce plants and animals to meet human needs. It is also a biological production process. This process depends on the growth and development of the selected plants and animals that are within the local environment.

1.5. AGRICULTURE AS ART, SCIENCE, AND BUSINESS OF CROP PRODUCTION

Agriculture is defined as the art and science and the business process that can be used for the production of the crops and the livestock for economic purposes. As an art, it is used to embrace the knowledge of the way that can be used to perform the operations of the farm in a skillful manner.

Here the skill can be classified or categorized as a physical skill. The physical skill involves the ability and capacity to carry out the operations in an efficient way. For example, handling of farms implements, animals, sowing seeds, application of the fertilizers and pesticides, etc.

Whereas, mental skills includes when the framers are able to take a decision that is based on the experience. These include:

- Time and method used for plowing;
- Selection of the crop and cropping system that can suit the soil and culture;
- Adopting the improved farms for practices, etc.

In terms of science, it is defined as it utilizes all the modern technologies that are developed on scientific principles. These principles include such as crop improvement/breeding, crop protection, economics, and crop production. These are being used to maximize profit with the yield.

For example, development of new crops and varieties by using the hybridization process, development of transgenic varieties that works as resistance to the pests and diseases, production of hybrids for each crop, use of a high variety of fertilizer that is more responsive, water management, use of bio-control agents that can be used to combat the pests and diseases, use of herbicides for controlling weeds, etc. (Figure 1.2).

Figure 1.2. Image representing agriculture crops in field.

Source: Image by Public Domain Pictures.

In terms of business, it is as long as agriculture is the way of life of the rural population. The production, in this case, is ultimately bound to the consumption. But it aims at the maximum net return. This return can take place through the management of land, water, capital, and labor.

It is also important in this field to employ the knowledge of the various sciences for the production of the food, feed, fire, and fuel. In recent years, agriculture is being commercialized to run as a business through the process of mechanization.

1.6. THE ROLE OF CERTIFIED ORGANIC FOOD AND FARMING

There are some of the farmers that have chosen to employ sustainable production practices. These practices will permit them to market their products as a "certified organics."

There are no such national estimates that can tell what percentage of production from the sustainable practices is being marketed in this way. But the figure is estimated at 30% (Figure 1.3).

Figure 1.3. Organic food farming.

Source: Image by Flickr.

The use of certified organic that is being labeled is developed as a way to assure the consumers that the food that is being consumed by them at the time of fast is being grown according to the practices. These practices are commonly associated with the word organic. Those who were involved in promoting the "organic" food have been seen what happened to the "Natural" food markets.

The process of certification is very useful in our food economy. It is important because consumers are usually not aware of the details of farmers whose product they are buying. It is being considered that there are many growers that are involved in the interprovincial and international trade.

There are some countries like Japan that do not give importance to the certification process as the importance for them is development. Japanese uses sustainable agriculture movement. This movement has instead focused on bringing the consumers and producers close together by simply creating consumer's producer cooperatives and buying groups. Thereby they are reducing the need for certification.

In this kind of system, the consumers may be even involved in the farm management decisions. This same approach is being used in a few more places in North America.

Certification process standards have been developed for most of the provinces present in Canada. But they all have a common base, they even similar in detail. These details depend on the ecological and economic conditions that exist in the region.

Based on "agroecological" principles, each set of standards are considered in a fact that commonly compromises between the ideal situation and the state of development that takes place in each region of sustainable practices.

For example, there are some standards that contain certain fertilizers and pesticides are permitted. There is a condition that may have detrimental effects on beneficial soil organisms, control agents, wildlife, and natural pests.

There are many cases where the understanding of the ecology of a pest or production of the system has not advanced to the point. The point where it can be assured the productivity and profitable system with the use of such products over the short term. As described above, the sustainable agriculture movement is very diverse. So, the sets of standards are also compromised between the different schools of the thoughts.

1.7. BRANCHES OF AGRICULTURE

The agriculture system has 3 main spheres: Geoponic, Aeroponic, and Hydroponic. Geoponic basically deals with the cultivation in the earth soil, aeroponic deals with the cultivation in air and Hydroponic deals with the cultivation in water.

As agriculture is the branch of science that encompasses the applied aspects of basic science. The applied aspects of agriculture science consist of the study of the field of crops and their management. Their management includes soil management.

1.7.1. Crop Production

Crop production deals with the production of various crops include food crops, fodder crops, fiber crops, sugar, oil seed, etc.

Crop production includes branches like pathology, agronomy, microbiology, soil science, entomology, etc. The main aim is to have better food production and the ways through which the disease can be controlled.

1.7.2. Horticulture

It's is the branch of agriculture that deals with the production of flowers, fruits, vegetables, spices, ornamental plants and condiments like narcotic crops-opium, etc., which are considered to contain some medicinal value and they are also used for beverages (Figure 1.4).

Figure 1.4. Cauliflower horticulture.

Source: Image by Wikimedia Commons.

1.7.3. Agricultural Engineering

It is considered as an important component that is used for the production and the horticulture. They are used particularly to provide the tools and implements. It aims to produce modified tools to facilitate proper animal husbandry and crop production tool, implements, and machinery that is used in animal production.

1.7.4. Forestry

Forestry deals with the production of large-scale cultivation of the perennial trees. These are used for supplying wood, rubber, timber, etc. They are also supplying raw material for the industries (Figure 1.5).

Figure 1.5. Image representing forestry development.

Source: Image by Needpix.com.

1.7.5. Animal Husbandry

The animal husbandry is when the animal is used for production; their maintenance is being done, etc. The maintenance is done of various types of livestock that are used for direct energy.

Husbandry is a common method that is used for both the crop and animals. The main objective that is used is to get maximum output by feeding, rearing, etc. The arrangement of crops is done in such a way to get the minimum requirement of the light and air. This arrangement is often known as geometry. The husbandry method is used for direct as well for indirect energy.

1.7.6. Fishery Science

Fishery management is basically for the marine fish and inland fishes which includes shrimp and prawns.

1.7.7. Home Science

Home science is the application and utilization of agriculture which is being produced in a better manner. When there is an enhancement in the utilization then there is an enhancement of production takes place. For example, when a crop is once used in the south, then it was noticed and found that it had many uses.

All these seven branches are in combination. Where the first three are being grouped for the crop production group, then the next two are being classified for the animal management and the last two are considered as allied agriculture branches.

In all these seven branches, the first three branches are being grouped for a crop production group, the next two groups are for the animal management, and the last two groups are allied agriculture branches.

1.8. FARMING TYPES

1.8.1. Subsistence Farming

Subsistence farming is done for the consumption of the farm owners that can be either primitive or intensive. This type of farming aims at fulfilling the needs of farmers and their families.

Primitive subsistence farming is considered as a type of subsistence farming that is done in a small area of land with traditional tools. These tools are like a hoe, digging sticks, Dao, etc.

This is considered as the most natural method that is being used for growing crops. This process takes place in a natural environment like heat, wind, condition of soil, and rain. These conditions support the growth of the crops.

Primitive farming further includes:

- **Shifting Cultivation:** This method is a primitive method where farmers clear the cultivated land. This process is done after harvesting the crops the lands are being burned. As a result, this helps in maintaining the fertility of the soil. This way the land is prepared for the next yield to get a good amount of yield.
- This method is known by a different names in different region. This method of shifting cultivation is being used by some countries like South America, South East Asia and India.
- **Nomadic Herding:** It is a kind of farming method that involves herders. In this method, the farmer travels from place to place with their flocks of animals. Herders are also used as a source of wool, hide, dairy product and meat that is being achieved from the livestock. This process is very common in India with the herders rearing sheep, yaks, camel, and goats.

Intensive subsistence farming is quite similar to primitive framing. The framers are practicing intensive farming on a wider area of land. They use modern machinery and tools and they also add chemical fertilizers to achieve better crops.

1.8.2. Commercial Farming

Commercial farming is a method where the farmers grow crops and rear animals for economic activity. As there is a need for a high amount of output, farmers use large areas for the cultivation process with the heavy use of machinery (Figure 1.6).

Figure 1.6. Image representing commercial farming.

Source: Image by Pikrepo.

Commercial farming is divided into three main categories:

- **Commercial Grain Farming:** This is the method where framers grow grains and use these grains for trading in the market.
- The most common crops that are used for commercial grain farming are wheat and maize. The farmers of Asia and Europe widely use the commercial grain farming method, and at the temperate grasslands of North America, this method is widely used.
- **Plantation Framing:** This method is a mix of agriculture and industry. This farming method is being practiced in a vast area of land. Plantation owners usually grow a single crop in the whole land. These crops can be banana, tea, coffee, etc. in a plantation. They use the technical support to process the crop on the farm or in a factory attached to it.
- The end product of the crop is used as a raw material in most of the industries. For example, the rubber is used by the rubber industries from the plantation that is being used as a raw material.
- **Mixed Farming:** It is a farming method that involves the cultivation of the crops, growing their fodder and rearing livestock. It is considered as a common method that is being practiced in the parts of the United States of America, Australia, and New Zealand, South Africa, and Europe. These countries use the mixed farming method for a living.

1.9. MAJOR CROPS USE FOR AGRICULTURE

As there are different, kinds of farming that are used to have different crops that are being grown by the framers. These crops are from staple grain to industrial crops and plants. These crops are being described below:

- **Rice:** It is the staple food crop that is being used by the whole world in a large quantity. Rice is grown in areas that usually contain high heat, humidity, and lots of rainfall. Around the world, the leading producer of rice is China which followed by India, Japan, Bangladesh, and Sri Lanka.
- **Wheat:** It is another staple food that is being used worldwide which requires moderate temperature. For the growth of wheat, there should be good rainfall availability during planting and it also required strong heat or sunshine while harvesting. Because

of this reason wheat is grown in the winter season in the northern states.

- **Maize:** Like wheat, maize also requires moderate rainfall and a good amount of sunshine at the time of harvesting. The countries growing Maize are Mexico, China, Brazil, Canada, North America, Russia, and India. These countries are primarily producing maize.
- **Millets:** This needs low rainfall and dry soil. These millets comprise grains like bajra, jowar, and finger millet. Millets are commonly grown in countries like China, Nigeria, and the dry areas of India.
- **Cotton:** It is a plantation crop that is considered as the main raw material used for the cotton textile industries. Cotton is best grown in the areas where the soil is black or alluvial. This is grown in the areas with low rainfall, bright sunshine and no snowfall. Countries that are producing cotton are Egypt, Pakistan, China, the USA, and India. These countries are the leading producers of cotton (Figure 1.7).

Figure 1.7. Cotton agricultural field.

Source: Image by Pixabay.

- **Jute:** It is also called as a golden crop. Jute is grown in the areas like tropical, where availability of the rainfall is high weather is humid. Jute is considered common in the coastal areas of Bangladesh and India.
- **Tea:** It is a plantation crop and considered an important beverage across Asia. Tea plantation is done in the sloping landscape where rainfall is provided even al around the year. The temperature in these areas is cool and not too high. The countries producing tea in the world are China, India, Sri Lanka, and Kenya (Figure 1.8).

Figure 1.8. Image representing tea plantation.

Source: Image by Flickr.

- **Coffee:** It is grown best in a warm and wet climate. Coffee is being grown on well-drained soil. Coffee is being produced by countries like Brazil, India, and Columbia. These countries are considered as the best producers of coffee in the world.

1.10. AGRICULTURAL SYSTEMS

Agriculture enterprises sue crops and livestock. As agriculture deals with the concept of labor supply, marketing, natural resource, genetic stock, finances, equipment, hazards, and nutrition. It is also possible to effectively manipulate each of these mechanisms of successful farming individually. It is a possibility that a better result can be obtained by simply treating these farming operations as a system.

The interactions may become more important than each component function by itself among the system component. Therefore, treating the production operations holistically is said to offer great management flexibility that provides more environmentally and economically sound practices. They even create safer and healthier conditions for the workers and as well as for the farm animals.

There are some research institutes that conducted research, extension, and education activities in the programs that are related directly and indirectly to the agricultural systems.

In the 19th and 20th centuries, U.S. agriculture was primarily based on maximizing production. This as a result becomes nationally as a breadbasket for the world. They are even generating large surpluses and even they are encouraging the farmers to remove some of the lands from the production.

These productions were successful in depressing the prices for many of the commodities. The prices that producers are receiving for the agricultural commodities are considered that they have not kept up with their increasing production cost.

Therefore, subsequently the narrow profit margins have been driven by the smaller producers out of agriculture. They are even forced the larger to become more efficient in their operations. These operations are being used for minimizing the expenses by maximizing the production.

Further, the public interest has been constrained by the farmers. This is done by increasing the demand for clean air and water, demand for the healthy soils, application of minimal chemical, treatment of humane animals.

Under this scenario, there are some smaller producers and producers are considered to contain specialty products that often struggle to succeed. The advantage of growing niche markets is being taken up by the specialty crop producers, organic, and horticulture producers. This allows them to retain more crop value from behind the farm gate.

There are few mechanisms that are driving many of research, education, and extension activities by the following programs:

- **Manure and Nutrient Management:** Manure is considered as a valuable, slow release fertilizer. This allows the farmers to recycle the animal waste back into crop production. When the animal feeding operations are larger and concentrated, the management of manure and nutrient becomes much more difficult.

If the waste is not being distributed to the farm fields, either readily or widely then the odors, nitrogen gases, and pathogens get accumulated. This program addresses the environmental issues that are being associated with manure management. As well as there are many beneficial that are using this manure for plant nutrition and to obtain useful by-products.

- **Organic Agriculture:** There are many consumers that are taking a greater interest in getting knowledge about where their food is coming from and how it is being grown. This is creating a demand for agricultural products that can adhere to the certain practice of healthy production. The organic operations seek to lower the production input and costs.

This is done to apply the environmentally sound practices which include natural manures, cultural pest management, and minimal soil disturbance to maintain healthy agro-ecosystem. These farms are believed to be smaller, closer to the consumers and tend to produce individually a variety of products.

The organic activities are believed to rely more on information that is site specific. The information that is found to be tied closely with the new precision technologies of farming.

- **Precision Farming:** These are the enabling technologies that are converging with agriculture and forestry. This is done to provide the measurement, storage, analysis, and helps in decision making which are the needs of these industries. The techniques are being developed to:
- To make precise measurements and continuously monitoring the field, product, forest conditions through sensors and controls.
- To organize the large volume of data with the spatially referenced database.
- To analyze and interpret the information using some decision support system. That allows the producers to make economically favorable choices.
- **Small Farms:** Agriculture when coupled with the relatively low prices for the farm products, basically encouraged many of the agricultural producers to farm more acres and raise a greater number of animals. Therefore, the presence of smaller farms to

enhance the quality of life in rural communities. It also helps in preserving the open green spaces.

It also provides locally produced fresh products, sustaining local businesses and it also creates the opportunity for the rural youth to develop. There are some programs that can increase the viability of the small farming culture and these are supporting education programs, programs accessing services, technical assistance, and other resources.

- **Sustainable Agriculture:** The rural community strength and prosperity are closely tied to agriculture sustainability. It is not that only the rural residents get benefits financially from agricultural enterprises. But they are believed to also enjoy living among the physical and social environment which is created by them only. Therefore, with all this maintain these latter amenities are found to be critical to the financial success and for the well-being of the communities.

1.11. APPROACH USED BY FARMING SYSTEM

1.11.1. Farm Household System

It is a group of people who are usually related, who individually or jointly helps in providing the management, labor, land, capital, and other inputs that are required for the production of crops and livestock. They are considered to consume at least a part of the farm produce.

1.11.2. Crop System

Crop system is an arrangement of the crop population that helps in transforming the solar energy, nutrients, water, and other inputs that are useful for the biomass. The crops can be of different species and different varieties. But they are considered to constitute one crop system if they are being managed as a single unit.

The crop system is considered as a subsystem of the cropping system. For example, if in the maize crop system, maize is the dominant crop which is grown in association with the other crops.

1.11.3. Cropping System

Cropping system is a land that is being used as a unit comprising the weeds, soils, crops, pathogens, and insect subsystem. They transform solar energy, nutrients, labor, water, and other inputs into feed, food, fuel, and fiber. The cropping system is considered as a subsystem or component of the farming system. It helps in representing the cropping pattern that is used on a farm.

It also represents their interaction with the farm resources, other farm enterprises and the available technology and environment that can be physical, biological, and can be sociological. This helps in determining their makeup.

Cropping pattern means the proportion of area that is under various crops at a point of time in a unit of area.

Cropping pattern indicates the yearly sequence and spatial arrangements of the crops that is being fallowed in an area. The crop sequence and crop rotation are generally used synonymously.

Crop rotation refers to the recurrent succession of the crops on the same piece of the land that is either in a year or over a longer period. The component of the crops is being chosen so that the soil health may not get impacted.

1.11.4. Livestock System

Livestock system is the method where the land is being used which comprises the pastures and herds and auxiliary feed sources that help in transforming the plant biomass into the animal product (Figure 1.9).

Figure 1.9. Livestock grazing on lands.

Source: Image by Flickr.

1.11.5. Herd System

Herd system deals with the arrangement of the domesticated animal population and their feed sources.

1.11.6. Agro-Ecosystem

Agro-ecosystem is the ecological system that are modified by human beings to produce the food, fiber or other agricultural products. Like, the ecological system the agro-ecosystem is often structurally and dynamically complex. But their complexity arises primarily from the interaction that takes place between the socio-economic and ecological processes.

1.11.7. Integrated Farming Systems

The integrated farming system means that when the farming system is integrated of crops, aquaculture, Agri-horticulture, livestock, and agro-forestry, etc. for the efficient use of farm resources to expand the farm income and sustainability in the production.

To distinguish between the two, another system is being coined which is called a specialized farming system. This specialized farming system is the one that concentrates basically on the single cropping system or sequence of farming enterprises. This includes animal breeding, dairying, etc.

This process is done to achieve the highest degree of precision management. This precision management takes place with the minimal diversion of resources or attention to diverse crops or enterprises. They differ widely in the managerial skills and also involve extremely different implements or capacity of handling the produce.

The integrated farming system is considered a component of farming system. That is being introduced to a change in the techniques of the farming. This is done basically to achieve the maximum production in the cropping pattern and to take care of the optimal utilization of the resource.

Farm waste that is collected is recycled later to achieve a productive purpose in the integrated farming system. It is not like the specialized farming system, as an integrated farming system is the activity that focuses on a few selected, interdependent, often interlinking, and interrelated production systems. This system is based on a few crops, animals, and related subsidiary professions (Jayanthi et al., 2002).

It is believed that the farming system approaches are not entirely a new approach for the farming communities. For the developing countries and

India it is not an exception. There are some insights into the traditional farming activist that indicates that the intensive farming system was in existence from the times of immemorial.

In a given farm, the farming activities may comprise of any one of these give below or a combination of the following:

- Apiculture (beekeeping);
- Cropping: mono or multiple;
- Horticulture: vegetable, orchards, plantations, flower cultivation or nursery Agroforestry systems which includes agri-horticulture, agri-pasture, silvi-pasture, agri-silvi-pasture, etc.);
- Fisheries;
- Sericulture;
- Sheep/goat rearing;
- Poultry;
- Livestock: milch animals/draught animals; and
- Piggery.

A farming system can comprise one or more than one enterprise. In most of the situation, only one of them is primarily important.

1.12. AGRICULTURE SYSTEM BY REGION

There are certain types of agriculture system regions, which are discussed below:

- Nomadic herding;
- Livestock ranching;
- Shifting cultivation;
- Rudimentary tillage;
- Intensive subsistence tillage (with paddy dominance);
- Intensive subsistence tillage (without paddy dominance);
- Commercial plantation;
- Mediterranean agriculture;
- Commercial grain farming;
- Commercial livestock and crop farming;
- Subsistence crop and livestock farming;

- Commercial dairy farming; and
- Specialized horticulture.

1.12.1. Nomadic Herding

Nomadic Herding system is an extensive farm of animals that are grazing on natural pasturage. This process takes place at constant on a seasonal migration of the nomads. Their flocks' nomadic handling is confined to relatively sparsely populated parts. These populated parts are of the world where the natural vegetation is mainly considered as grass.

Nomadic herding is mainly concentrated or present in the Saharan Africa areas includes Libya, Mauritania, Mali, Niger, Sudan, Chad, and Algeria. It is also present in the central parts of the Asia of Scandinavian countries that include areas of Norway, Sweden, and Finland and also in Northern Canada.

There are some characteristics that are explained below:

- The main characteristic of nomadic herding are the continued movement of people with their livestock in search of forage for the animals.
- This type of system is carried mainly to produce food for the family to fulfill the needs of clothing shelter recreation.
- The Bedouin of Saudi Arabia the Tuareg of the Sahara also practice nomadic herding in the desert semi desert areas of North Africa southwest Asia.
- Nomadic herding is considered as an ecological or rear ecological system of agriculture.
- It is a declining type of agriculture that continues to become less important.
- The chief characteristics of nomadic herding are described below:
- Seasonal pattern of movement;
- Transhumance; and
- Many kinds of animals grazed.

1.12.2. Livestock Ranching

Livestock Ranching is extensive temperate grassland that is once named by the nomadic herdsmen or by the hunters. These hunters are found as permanent Ryle where a large number of cattle, sheep, goats, and horses are kept.

Livestock ranching is present in the areas of America, Australia, Brazil, Peru, Argentina, New Zealand, and Republic of South Africa. These areas are the nearest equivalent to the nomadic herding.

Livestock ranching has some characteristics that are explained below:

- Tim livestock ranchers have a fixed place of residence and operate as individuals rather than within a tribal organization.
- The livestock ranchers specialize in animal husbanding to the exclusion of crop raising even though both live in arid on the semi-arid region.
- Livestock ranching differs farm nomadic herding in time important aspects.
- There is little or no migration;
- The vegetation cover is continuous;
- The animal's arc rose for sale;
- Ranches one scientifically managed; and
- Commercial grazing supports the development of the town's communications.

1.12.3. Shifting Cultivation

Shifting cultivation is the essentially land rotation system. In this system, farmers are using machetes or other bladed instruments. These instruments are used to chop away the undergrowth that is present in the small pat etches of the land.

After this, they kill the trees by cutting off the strips of the bark completely around the truck. These cleaning techniques have gluten the shifting cultivation.

The shifting cultivation is the primitive form of soil which is being utilized usually in a tropical rainforest. This cultivation process is being followed by the tropical lowland hills areas that are present in the center of America, Southeast Asia, Africa, and Indonesia (Figure 1.10).

Figure 1.10. Tropical rainforest.

Source: Image by Flickr.

There is a certain characteristic of the shifting cultivation which is explained below:

- The shifting cultivations grow food grains like rice, maize, millet, jobs, beans, vegetables, and soybean.
- Shifting cultivation has been described as an economy of which the main character is the rotation of fields rather than rotation of crops.
- The farmer grows food only for their family in these agriculture systems. Some small surplus if any are exchanged on bartered (exchange of a commodity for a commodity) or sold for the crash in the neigh boning markets.
- Shifting cultivation is called by a different name in different parts of the world. It is generally known as a slash burn and bush fallow agriculture. It is variously termed as ladang in Indonesia, Milpa in Central America, and Mohole in the Congo central Africa.
- In the hill tracts of north-Cast India Thinning is the dominant economic activity. Our 86% of the people living on hill arc dependent on shifting cultivation.

1.12.4. Rudimentary Tillage

Rudimentary tillage is considered as the most confined process to the tropical lands. This type of system is present in the Central and South American areas, Africa, and SouthEast Asia.

There are certain characteristics of Rudimentary tillage those are explained below:

- Crop rotation occurs mostly then the field rotation.
- In this area's potatoes, maize, sorghum, sweet potatoes and banana are grown.

1.12.5. Intensive Subsistence Tillage

The intensive subsistence tillage is the form of agriculture process which is best developed in the partially confined monsoon lands of Asia.

- The intensive subsistence tillage is being dominated by the paddy. This type of process is being practiced mostly in tropical Asia. It is carried by China, Japan, India, Myanmar, Sri Lanka, Bangladesh, Malaysia, and Philippines, etc.
- There are some characteristics of the intensive subsistence tillage that are explained below:
- This process includes, where only one crop of paddy can be raised. The fields are normally used in the dry season to raise other food or cash crop such as sugar tobacco on oil seeds on the fiber crop jute.
- Farming is also considered as intensive that double or treble cropping is practiced. That is several crops that are grown during a year on the same land.
- Asian farmers are now producing even greater yields per acre. This is because to improve the varieties of hybrid rice by the recent introduction.

1.12.6. Commercial Plantation

The commercial plantation is the process that is specialized as commercial cultivation that includes cash or estates or plantation. This is considered as a very distinctive type of tropical agriculture that is found in many parts of the world.

This term plantation agriculture was originally applied specifically to the British settlements. This process takes place in the regions of America, southeast Asia and west India. The commercial plantation is mainly cultivated by the people who belong to Negro or other colored labor.

There are certain characteristics of the commercial plantation that are explained below:

- Some of the main plantation crops are rubber, oil palm, cotton, copra, beverages like coffee, tea, cocoa, fruits like pineapples, bananas, as well as sugar-cane jute.
- The plantation forms arc generally large and arc found mainly in the thinly populated areas. The size of the farm varies from 40 hectares in Malaya India, too 60,000 hectares in Liberia. In these estates on large disciplined but unskilled labor, force is necessary.
- A plantation is a landholding that is devolved to the specialized production of one tropical or subtropical crop that is raised for the market.
- The continent wise analysis reveals that Asia is the leading producer of Jute (96%), rubber (90%), tea (87%), coconut (37%), and tobacco (46%) of the total world production Asia share in the production of sugarcane is 39% that of banana, oil palm is 25% each (Hussain, 1996).
- The climatic hazards include strong winds, topography, and drainage soil vegetation condition. These climate hazards are barriers or may even prevent the development establishment of the plantation.
- Accessibility for the connectivity availability of labor difficulties of clearing vegetation prevalence of discovering sin sect pests, weeds, rapid deterioration of the tropical soil, soil erosion is some of the main problems of plantation agriculture.
- There are certain characteristics features of the commercial plantation that may be summarized as follows:
- Estate farming;
- Foreign ownership local labor; and
- Fanning in estates is scientifically managed.

1.12.7. Mediterranean Agriculture

Mediterranean Agriculture takes place within the Mediterranean climatic region. This process is done in a region where there is winter rain, summer drought and the evolvement of a distinctive type of agriculture is present.

This type of agriculture typology is being confined to the coastal areas. This type of agriculture process is done in the Mediterranean Sea in Europe, North Africa coastal strip, Asia, and minors.

This system is also found outside the Mediterranean Sea in California (USA). It also practiced in the central Chile of the Southeast of Cape provision and Southwest of Australia.

There are some characteristics of the Mediterranean Sea that are discussed below:

- The agricultural landscape of the Mediterranean region has been largely affected by long day summers occurrence of rains which occurs during the winter season devices for artificial irrigation during drought periods of summers.
- This type of farming is also found in irrigated semi-descent areas in similar latitudes.
- In recent times, the farmer has begun using irrigation ill that is one of the major ways which has led to the expansion of crops such as the citrus fruits.
- Traditional Mediterranean agricultural is bored on what barely cultivation in the rainy wastes season raising drought resistant vine tree crops like the grape olive, fig, small livestock herding particularly of sheep, goat, and pigs.
- The Mediterranean land, in fact, the Orchard lands of the world the heart of the world's wine industry.

1.12.8. Commercial Grain Farming

Commercial Grain farming is another market-oriented type of agriculture. In this type of system, farmers are specialized in growing wheat variety on a less frequently that rice and corn.

Commercial grain farming is a great wheat belt that gets stretch through Australia to the plains of the interior of North America. It is also practiced in the steppes of Russia, the pampas of Argentina and together with the United States and Canada farmers with soviet unions.

There are certain characteristics of the commercial grain farming system that is explained below:

- Widespread use of machinery enables commercial grain farmers to operate on this large scale indeed planting that tells the harvesting grain is more completely mechanized than any other form of agriculture.
- The commercial vain fanning is extensive. The main characteristics of these systems are:

i. Low use of irrigation fertilizer;
ii. Big farm size;
iii. Long distance of farm from the market;
iv. Low production rate; and
v. Comprehensive use of heavy machines.

- Wheat is the main crop; Mize, Barleys, oat ore other important crop. The wheat production regions are divided into two belts.

i. Winter wheat belt; and
ii. Spring wheat belt.

1.12.9. Commercial Livestock and Crop Farming

This type of farming system is found throughout Europe from the land in the west through central Europe to Russia. It is also found in the North America cast at around 98° meridian in the pampas of Argentina, southeast Australia, South Africa, and New Zealand.

There are certain characteristics of commercial livestock and crop farming that is being discussed below:

- Mixed farming is essentially associated with the density population. Urbanized industrialized societies dependent upon high incomes for the sale or its products upon the manufacturers industry for the provision of its inputs.
- The main characteristic of mixed farming is that farms produce both crops and livestock.
- In mixed farming, a number of crops are grown. Crenels dominate the cropland, by using the leading grain varying with climate and soil. A large portion of cereals is fed to animals on farms. Livestock feed on the crop grown on the farm graze the post wire.
- Mixed farming is characterized by the high expenditure on

machinery, farm building, extensive use of machinery arm, buildings fertilizers, and also by the skill experts of farmers who need to know about all aspects of farming to grow the market range of their product successfully.

1.12.10. Subsistence Crop and Livestock Farming

Subsistence crop and Livestock farming are operated in the areas of Northern Europe, Mountain region of Mexico and the Middle East.

There are some characteristics of subsistence crop and livestock farming which are explained below:

- Traditional way of farming;
- Produced crops and raised livestock mainly used for own subsistence;
- Wheat, maize, rye, barley etc. are the main crops;
- Seeds are poor quality and animals are poorly husbanded;
- Sheep and goats are the most important animal; and
- Capital input is normally unknown.

1.12.11. Commercial Dairy Farming

Commercial dairy farming is the process of rearing of cattle for milk and its product like butter, cheese; dried milk, etc. are known as dairy farming.

This type of farming process is being practiced in the area of Europe, Northern USA, Canada, New Zealand, Netherland, Belgium, France, Switzerland, Australia, Finland, and Denmark. It contributes about 40% of the agricultural income.

There are some characteristics of commercial dairy farming that are explained below:

- Nearly 80% of the total milk production of the world is produced in Europe, Russia, Anglo America, Australia, New Zealand (Hussain, 1996).
- Dairying is capital intensive farming. A modern dairy farming needs huge amounts of development of infrastructural facilities capital is required for the punch of mechanical equipments like milking machines milk freezers, feeding towers, born silos for the storage of fodder in winters.

- Modern method of daily farming cattle breeding herd management allows high yields of dairy products. A cow in temperate latitudes under normal healthy conditions yields as much as 3000 kg of milk per year.
- The size of cattle in dairies varies from country to country from farm to farm depending on the size of the holding. In the United Kingdom, for example, the ratio of cattle pasture is one cow after one acre. The average size of dairy cattle in northwest Europe is only five cows per farm (Table 1.1).

Table 1.1. The Countywide Position in Some the Denying Countries of Milk per Cow per Annum

Country	Amount (kg)
USA	3767
Japan	4284
Canada	2896
UK	3797
Denmark	3946
Netherlands	4202

Source: ResearchGate.net.

It can see from the data above that the average production of the milk per cow per annum varies between 2897 kg in Canada to 4284 kg in Japan.

1.12.13. Specialized Horticulture

Specialized horticulture is the vegetation of vegetables, fruits, and flowers, which is commonly known as horticulture.

This type of agriculture process is being practiced in a well-developed and densely populated industrial area. These areas are North West Europe, Denmark, France, Germany, Italy, and Britain.

There are some characteristics that are explained below:

- The important vegetation region is California Rio Ground boring of Texas Florida Netherlands, Rhone valley etc. fruits regions are west of Paris, Rhine valley lake region of Switzerland, Mendoza, San Juan of Argentina etc. (grapes production) south western Germany (apple).

- The market gardens are scientifically managed to achieve optimum yields and to have good amount returns.
- In horticultural the farms are small. Such farms are located where communication links and the consumption centers are appreciably good. The land fruits and vegetable gardening are very intensively cultivated. Soil fertilizer work is done by hand labor.
- Besides south Arab, Iraq is famous for dates. India, South-East Asia are famous for spices, pineapple, mango etc.

The following types of agricultural system regions:

- Nomadic herding;
- Intensive subsistence tillage (with paddy dominance);
- Livestock ranching;
- Shifting cultivation;
- Rudimentary tillage;
- Intensive subsistence tillage (without paddy dominance);
- Commercial plantation;
- Mediterranean agriculture;
- Commercial grain farming;
- Commercial livestock and crop farming; and
- Subsistence crop and livestock farming.

1.13. CONCLUSION

This chapter is about Agriculture's definition, history about its introduction in the universe. As the plants were the first to arrive. From which around 50,000 species of plants are still available. After plants, then animal species arrived. Many of the species are said to disappear with time.

Agriculture is a combination of art and science. It is considered as a complete process which includes input, processing, and output processes. Where input is the seed that is to be grown, processing includes fertilizers, manure, etc. and output is the product that is received after harvesting.

Agriculture contains certain branches like crop production that involves growing of crop like food, fodder, fiber, oil, etc. Horticulture includes the production of fruits, flowers. Similarly, there are many other branches of agriculture that are present like forestry, animal husbandry, fisheries science, home science, etc.

The farming system can be defined as two types, namely subsistence farming and commercial farming. Where subsistence farming is farming that is done on a small scale for fulfilling the needs of the family. Whereas commercial farming is done for benefiting commercially. Which includes the growing of crops and the rearing of animals for selling purpose.

So, this chapter explains all the agriculture methods or systems that are used now a days for feeding people and fulfilling their requirements. As agriculture is an important part of everyone's life, so it is also important to maintain that culture and practices.

REFERENCES

1. Agriculture and Farming Systems, (n.d.). [eBook] Available at: http://www.fao.org/3/a0416e/a0416e03.pdf (accessed on 2 April 2020).
2. An Introduction to Agriculture and Agronomy, (n.d.). [eBook] Available at: https://www.mobt3ath.com/uplode/book/book-7871.pdf (accessed on 2 April 2020).
3. Eap.mcgill.ca. (1989). An Introduction to Sustainable Agriculture. [online] Available at: https://eap.mcgill.ca/publications/EAP16.htm (accessed on 2 April 2020).
4. Edwards, C., (2001). Agricultural systems: Ecology. Encyclopedia of Life Sciences.
5. Hano, A., (2011). Introduction to Agriculture. [online] ResearchGate. Available at: https://www.researchgate.net/publication/282905517_Introduction_to_Agriculture (accessed on 2 April 2020).
6. Hussain, M., (2016). "Whittlesey Classification of World Agricultural Region: A Conceptual Study." ResearchGate. [online] Available at: https://www.researchgate.net/publication/305392168_Whittlesey_Classification_of_World_Agricultural_Region_A_conceptual_study (accessed on 2 April 2020).
7. Nifa.usda.gov. (n.d.). Agricultural Systems | National Institute of Food and Agriculture. [online] Available at: https://nifa.usda.gov/topic/agricultural-systems (accessed on 2 April 2020).
8. Rana, S., (2011). Farming Systems and Sustainable Agriculture. [online] ResearchGate. Available at: https://www.researchgate.net/publication/309211183_Farming_Systems_and_Sustainable_Agriculture (accessed on 2 April 2020).
9. Toppr-guides, (n.d.). Introduction to Agriculture: Subsistence, Commercial Farming, Examples. [online] Available at: https://www.toppr.com/guides/geography/agriculture/introduction-to-agriculture/ (accessed on 2 April 2020).

CHAPTER 2

Science of Agronomy

CONTENTS

This chapter named 'Science of Agronomy' includes a basic introduction about agronomy. There are several definitions provided by different organizations and different researchers. It can be said that agronomy is the application of the science of technology in the context of agriculture.

In the first half of the chapter, it provides insight into what agronomy is and who is an agronomist. This also provided different definitions of agronomy. In this chapter, the readers will also get to know about the history of agronomy and how agriculture responds to technology.

In the other half of the chapter, readers will get to learn about the scope of agronomy and also the principles followed by agronomy. In the last, chapter, some of the sub disciplines of agronomy have been mentioned such as plant breeding, biotechnology, soil science, soil conservation, and agroecology.

2.1. INTRODUCTION

A science that helps in feeding the world is referred to as agronomy. The agronomy can be considered to be the backbone of all agricultural sciences. The reason behind this is the management of soil and water which is exclusively an agronomic domain, intending to achieve the production potential of high yielding varieties, in a green revolution.

It is required for the agronomists that they can be able to synthesize production practices from a number of different fields of specialization. The problem of global food security remains still unsolved.

Along with the increasing population, demand for food in the world also increases, whereas there are important factors in production such as cultivated land and fresh water which are continuously decreasing.

There are current trends in world agriculture which shows that it is imperative to find a scientific as well as a rational way for developing it. This is a way that can steadily increase the output and also, it makes sure about the sustainable use of resources in the process for the promotion of agricultural development.

The word 'agronomy' has been derived from two Greek words, 'agros' and 'nomos' which means field and to manage. Therefore, the word agronomy means 'the art of managing field.' In technical terms, agronomy can be referred to as the 'science and economics of crop production by the management of farm land.'

Agronomy is considered as the art and the underlying science in production as well as improvement of field crops along with the efficient use of soil fertility, water, laborer, and some other factors which are related to crop production.

Agronomy is said to be the field of study and practice of ways and means of production of food, feed, and fiber crops. A branch of agricultural science which deals with the principles and practices of field crop production, as well as management of soil for higher productivity, can be referred to as agronomy.

Considering all the branches of agriculture, agronomy is said to occupy a pivotal position and so, is regarded as the mother branch or the primary branch. Similar to agriculture, agronomy is also an integrated as well as an applied aspect of different disciplines of pure science (Figure 2.1).

There are many clear branches of agronomy which include:

- Plant breeding;
- Biotechnology;
- Soil science;
- Soil conservation; and
- Agroecology.

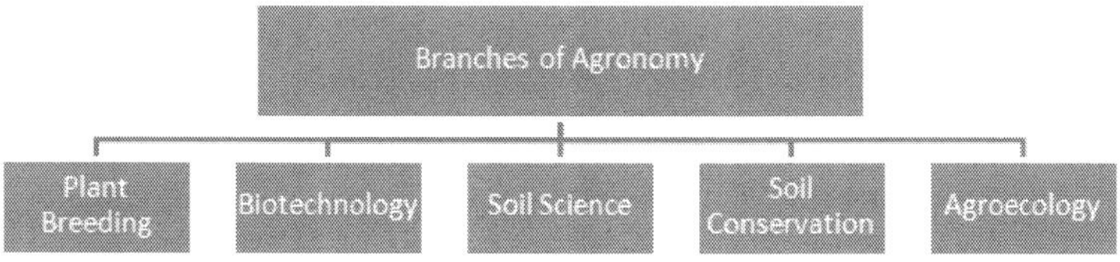

Figure 2.1. Branches of agronomy.

2.2. WHAT IS AGRONOMY?

- Agronomy can be defined as the application of science and technology from the fields of biology, chemistry, ecology, economics, water science, soil science, pest management and genetics to the improvement as well as management of the major food crops of the world.

- Agronomy can be defined as differentiated from the plant sciences of horticulture, range science, pomology, and some other applied plant sciences by particular crops considered and the scope.
- Agronomic crops are those crops that occupy large acreage and so, are the bases of the food and fiber production systems of the world and that are mechanized most of the time.
- There are some examples that include wheat, corn, rice, soybean, alfalfa, and forage crops, sugar beets, beans, canola, and cotton. These are also referred to as 'field crops.' Typically, these crops contain a large majority of agricultural acreage and crop revenue of a country.
- Agronomy is said to be a science that looks at agriculture from an integrated as well as holistic perspective, further, informing practical decisions which pertain to the food and fiber production.
- The whole range of influences is considered by agronomy that is on crop production. This includes climate and adaptation, soil, crop genetics, water, and water availability, the properties of the soil and how the soil interacts with the growing crop; what nutrients are required by the crop; the ways in which the crops grow as well as develop; and how best to control weeds, fungi, insects, and some other crop pests.
- There are non-biological considerations which include economic requirements and consumer and farmer behavior, which also inform agronomic practices, as do environmental constraints.
- Agronomy is referred to as science which considers how to grow crops in an effective and profitable manner while conserving natural resources and also, protecting the environment.
- There is an integration of various science which is required by the agronomy. It does require collaborations among a number of diverse fields, which include soil, plant, and weed science. There are many disciplines also which include ecology, climatology, entomology, and economics.

There are some effective crop production methods and these methods are grounded in scientific research and therefore, nature is continuously evolving and improving.

2.2.1. What Is an Agronomist?

Agronomists are the scientists who work with the plant and soil and further develop the innovative farm practices and technologies to boost the crop yields, improving the farm profitability and sustainability and furthermore, protect the environment.

Most of the time, agronomists are specialized in areas such as irrigation or water science, soil fertility, plant physiology, plant breeding, economic, crop management and pest control. They have the capability of addressing and an integration all of the areas which influence crop production.

Agronomists can also act as professional practitioners, advisers, and educators. They tend to work directly with the farmers, companies, and some others in the agricultural community to implement the latest methods and tools, which can be used for growing crops in a profitable as well as sustainable manner.

An important role is played by the agronomists in the issues of global concern, which include food and water security, air quality and climate change, soil, and water conservation, health, and nutrition, rural communities and economic growth, wildlife habitat in agriculture and many other areas of public concern.

2.2.2. What Is Agronomy vs. Horticulture: Distinctions Not Well Settled

From the outset, it is important to realize the distinction between agronomy and horticulture which is not well established. Both terms can also be used to the same crop.

This is similar for the grain or forage corn (agronomic) and sweet or young corn (horticultural) and also, with pineapple grown for fiber (agronomic) and those grown for fruit (horticultural).

Furthermore, the primary usage is not the only basis of distinction. There is some other certain global development which may cause a significant change. This change will affect the traditional uses of some of the specific crops.

Let's take an example. There can be an increase in the world market demand for coconut water from young nuts which eventually can have drastic effects on the production of mature nuts for copra. If in case, this happens significantly, then the alteration of the established classification of coconut as an agronomic crop can take place.

Along with this, there are other criteria of classification which include the intensity of care and value of the product that are also considered. These can vary with time as well as place.

Let us consider the case of the Philippines. There, the sweet potato and white potato along with the field beans are considered to be the horticultural crops based on their intensive management. Nevertheless, according to Lantican (2002), they are classified as agronomic crops in some of the countries where they are grown by using the mechanized system on an extensive scale.

As per the view presented by Janick (2002), there are present divisions of agriculture that are into agronomy and horticulture along with forestry and this originated from medieval agriculture in Europe. A system of differentiation was used based on crops and intensity of production.

If there is a production system that is in open fields or meadows are used to produce the grains and fodder for animals, then this can be considered into agronomy. Horticulture is something that is involved with the production of tree fruits and vines, ornamental plants, vegetables, herbs, and medicinal plants in the kitchen gardens.

According to some distinguished author, this system breaks down in the tropics and here it is never clear where the ending of agronomy is and where it is the starting of horticulture.

2.3. HOW AGRICULTURE RESPONDS TO THE WORLD AROUND US

As the definition says, agronomy is the study of the ways in which the plants and soil can be best used for food, fuel, and fiber and reclamation.

The agronomy is considered to be important and its important combines the studies of ecology, chemistry, biology, earth science, and genetics to determine the capabilities of several different types of plants as well as soils in their environments.

The importance of agronomy provides the farmers with agricultural information about the way for growing and caring for the plants as well as soils in some specified environments.

There are some factors such as climate, roots, moisture, weeds, fungi, pests, and erosion which can pose considerable challenges when the farmers attempt to produce a plentiful harvest. Agronomists tend to study these agricultural hurdles so as to discover the ways of integrating crops into the environment in many ways that will help them in prospering (Figure 2.2).

Figure 2.2. Agronomist and farmer inspecting weed.

Source: Image by Flickr.

Soil and plant samples are tested by the agronomists from all over the world to better understand the interaction between their properties and genetic make-up along with their surrounding environment.

Then, they take this research for developing some innovative farm practices as well as technologies that will boost crop yields and protect them against weeds and pests.

According to the findings of an agronomist, they provide farmers, present across the globe, with the important information and tips that are related to producing crops and also, for conserving our natural environment.

Agronomists are motivated for exploring the agricultural resources continuously by their concerns related to food and water security, soil loss and degradation, air quality and climate change, health, and nutrition and many others.

It is possible that the farmers will integrate every new agronomy finding along with their approach to tending to their crops. This is done so as to preserve, expand, and reclaim the use of plants as well as soil for food and fuel.

The human can allow the plants and soil so as to reach their full potential ad can reap the benefits which can be done by understanding the importance of agronomy and the agricultural resources around us.

2.4. SCOPE OF AGRONOMY

Agronomy is considered as a dynamic discipline. Along with the advancement of knowledge as well as a better understanding of plant and environment, there are agricultural practices that are modified or even new practices that are developed for high productivity.

Let's take an example, the availability of herbicides for controlling the weed can result in development of a vast knowledge about the selectivity, time, and method of application of herbicides.

There are gigantic irrigation projects which are constructed to provide the irrigation facilities. Nevertheless, such kind of projects has created side effects such as water logging and salinity.

Proper water management practices are developed to overcome this problem. There is increase in the population pressure but the area under cultivation is static. Therefore, a greater number of crops are required to be grown on the same piece of land in a year.

Intensive cropping has come into vogue as a result of this. Similarly, there are no tillage practices that have come in place of clean cultivation which is considered to be the result of increasing cost energy. Similarly, it is important to develop new technology to exploit its full potential.

There are certain factors that have restricted the increased agricultural production and these factors include low soil fertility, crop varieties of low genetic yield potential, poor agronomic practices, non-availability of production inputs, inadequate control of disease and insects, government economic policies affecting agriculture and weak research and extension programs.

There are some agronomic practices which can make the finite land water resources much more productive and these practices include the restoration of soil fertility, use of proper seed rates, preparation of good seedbed, proper conservation, and management of soil moisture, correct dates of sowing for each improved variety and proper control of weeds.

A principle can be referred to as a scientific law, which helps in explaining the natural action. The agronomic principles can be defined as the ways and means which helps in better management of soil, plants, and environment for economically maximum returns per unit area.

The principles of crop management majorly depend upon the type of farming which includes diversified, specialized, mixed, and integrated and also, depends on the physical as well as technological available, dry farming, irrigated farming, and rain fed farming.

There are some fundamental principles of agronomy which can include:

- Choice of varieties of a crop which are adaptable to the particular agro-climate, land situation, soil fertility, season, and method which can be used for cultivation and befitting to the cropping system.
- Planning, programming, and executing measures for maximum utilization of land, capital, labor, sunshine, transport, temperature, rain-water, humidity, and marketing facilities.
- Proper field management can be done by tillage, checking soil erosion, preparing field channels and bunds for irrigation and drainage, leveling, and adopting other suitable land improvement practices.
- It is important to timely apply properly as well as balanced nutrients to the crop or crops in sequence and further, improve the soil fertility and productivity. Correction of bad-effects of soil reactions and conditions should be done timely and further, increasing soil organic matter by applying the green manure, organic wastes, farm yard manure, bio fertilizers, and profitable recycling of organic wastes.
- Adoption of multiple cropping along with mixed or intercropping to ensure harvest even under adverse environmental conditions.
- Choice of quality seed or seed material and maintenance of requisite plant density per unit area along with, healthy, and uniform seedlings.
- Adoption of adequate, need-based, timely, and proper plant protection measures that can use to prevent weeds, insect-pests, pathogens, as well as climatic hazards and correction of deficiencies and disorders.

- Proper water management to crop, soil, and environment through conservation and utilization of soil moisture as well as by water, that is available in excess, needs to be done.
- Adoption of suitable and proper management practices includes intercultural operations to get maximum benefit.
- Adoption of suitable post-harvest technologies should be done.
- Adoption of suitable method as well as time of harvesting of crop to reduce field-damage and to release land for succeeding crops and efficient utilization of residual moisture, plant nutrients and some other management practices.

2.5. SUB DISCIPLINES OF AGRONOMY

2.5.1. Plant Breeding

Plant breeding is the area of agronomy which includes selective breeding of plants to produce the best crops under various conditions. Plant breeding has increased the crop yields along with the reconstruction of new genotypes and it has improved the nutritional value of numerous crops (Figure 2.3).

Figure 2.3. Plant breeding of tomato.

Source: Image by Pixabay.

Also, it has resulted in the development of new types of plants. Let's take an example. There is a hybrid grain known as triticale which was produced by crossbreeding rye and wheat. Triticale has much more usable protein in comparison to wither rye or wheat.

Considering the fruit and vegetable production research, agronomy has been considered as an instrumental. It is understood that the role of agronomists includes referring to whether produce from a field which meets the following conditions:

- Land and water access;
- Market access;
- Risk protection (insurance);
- Quality and quantity of inputs; and
- Agricultural credit.

Plant breeding does apply the genetic principles to produce the plants which are much more useful to humans.

This can be accomplished by selecting the plants which are found to be economically or aesthetically desirable. This can be done by first controlling the mating of selected individuals and then can be done by selecting some specific individuals among the progeny.

There are many such processes that have repeated over many generations. These can change the hereditary makeup and value of a plant population that should be far beyond the natural limits of the population is exists earlier.

Plant breeding is said to be an ancient activity or dating which is from the very beginnings of agriculture. Possibly, soon after the domestications of cereal grains which was present in the earliest times, humans started to recognize the degrees of excellence among the plants in their fields.

Also, humans have saved seed from the best to plant new crops. There were such tentative selective methods that were considered as the forerunners of early procedures for plant breeding.

The results formed from the early procedures of plant breeding were conspicuous. There are varieties present today which are so modified from their wild progenitors that they are not even able to survive in nature. In some of the cases, the cultivated forms are different from the existing wild relative. It is so different that it is much difficult to even identify their ancestors.

These are the remarkable transformations that were accomplished by early plant breeders in a very short span of time, considering the evolutionary point of view. The rate of change was also probably greater in comparison to any other evolutionary event.

In the period in the mid-1800s, Gregor Mendel has highlighted the principles of heredity which was done by using pea plants. Furthermore, he provided the framework required for scientific plant breeding.

In the early 20th century, the laws of genetic inheritance were delineated. Along with this, there was a beginning towards applying them for the improvement of plants.

There are many major facts that emerged during the short history of scientific breeding. One of them is then an enormous wealth of genetic variability does exist in the plants of the world and also, that there is only a start that has been made while tapping its potential.

2.5.2. Biotechnology

Biotechnology is used by the agronomists for expediting and extending the development of desired characteristics considering the new varieties. Most of the time, biotechnology is considered to be a lab activity that requires field testing of the new crop genotypes which have been developed.

Along with the increasing crop yields, improving health and taste of foods, reducing crop vulnerability to environmental stresses and further, reducing the need for field applied chemicals, the application of agronomic biotechnology is increasing. This is increasingly being used for the novel uses other than food.

An example can be oilseed. Presently, the oilseed is used majorly for the margarine and some other food oils. This can also be modified to produce fatty acids for detergents, substitute fuels, and petrochemicals.

Biotechnology can be referred to as the use of biology for solving the problems and make some useful products. There is an area that is much prominent in biotechnology and that is the production of therapeutic proteins and some other drugs by using genetic engineering.

Biological processes have been harnessed by the people to improve their quality of life considering the 10,000 years, starting with the first agricultural communities. Approximately 6,000 years ago, the human started to tap the biological processes of microorganisms for making bread, alcoholic beverages, and cheese and for preserving the dairy products.

For the time more than a decade, the recombinant DNA or genetic engineering has dominated the biotechnology industry. This technique includes splicing the gene for a useful protein (most of the time it is a human protein) into production cells like yeast, bacteria, or mammalian cells in culture, then this starts to produce the protein in volume.

There is a new organism that is created while splicing a gene into a production cell. First of all, the biotechnology investors and researchers were not certain about whether the courts will allow them to acquire the patents on organisms. The reason was the patents were not allowed on new organisms that happened to be discovered and identified in nature.

But in the year 1980, the US Supreme Court (in the case of Diamond v. Chakraborty) solved the matter. They ruled out that 'a live human-made microorganism is the patentable subject matter.' This was the decision that spawned a wave of new biotechnology films and the first investment boom of the infant industry.

The recombinant insulin was the first product that was made in the year 1982 by using the genetic engineering to secure approval from the US Food and Drug Administration (FDA).

After that time, there were dozens of protein medications were genetically engineered and they all have been commercialized around the world. This includes recombinant versions of growth hormone, protein to stimulate the production of red and white blood cells, clotting factors, interferons, and clot-dissolving agents.

In the early years, there was a main achievement of biotechnology and that was the ability to produce naturally occurring therapeutic molecules in larger quantities than those which could be derived from conventional sources such as plasma, animal organs and human cadavers. Also, the recombinant proteins are not much likely to be contaminated with the pathogens or to provoke allergic reactions.

In today's era, biotechnology researchers are trying to discover the root molecular causes of disease and also, to intervene precisely at such a level. Most of the time, this means producing the therapeutic proteins which augment the own supplies of the body or that make-up for the deficiencies relate to genes, similar to the first generation of biotech medications.

The expansion of research has taken place by the biotechnology industry into the development of traditional pharmaceuticals and monoclonal antibodies which stop the progress of a disease. There are steps that are uncovered by the painstaking study of the gene (that is referred to as genomics), the proteins which they encode (which is referred to as proteomics) and the larger biological pathways in which they tend to act.

Along with the tools mentioned above, biotechnology has also involved merging biological information along with computer technology (which is referred to as bioinformatics). Possibly by applying the techniques of stem cell research and cloning, the replacement of the dead and defective cells and tissues (which is referred to as regenerative medicine) can be done and the use of microscopic equipment can be explored which can enter the human body (this is referred to as nanotechnology).

These disparate technologies have been integrated by the companies and academic laboratories in an effort to perform the analysis downward into molecule and also, in for synthesizing upward from molecular biology toward chemical pathways, tissues, and organs.

This is being used in health care. Along with this, biotechnology has proved to be much useful in refining the industrial processes by discovering and producing the biological enzymes which spark chemical reactions (catalysts).

This is helpful for environmental cleanup, using the enzymes which digest contaminants into chemicals that are not harmful and then die after consuming the food supply available to them. This is also useful in agricultural production by using genetic engineering.

The agricultural application of biotechnology is considered to be the most controversial. There are some activists and consumer groups which have called for bans on Genetically Modified Organisms referred to as GMOs. They have also called for labeling the laws to inform the consumers about the growing presence of GMOs in the food supply.

Let us consider the case of the United States. In the U.S., the introduction of GMOs into agriculture started in the year 1993. This was the time when the FDA approved the Bovine Somatotropin (BST). BST is a growth hormone that helps in boosting milk production in the dairy cows. In the next year, the FDA approved the whole food which was genetically modified which is a tomato that was engineered for longer shelf life.

From that time, in the United States, Europe, and many other places, the regulatory approval has been won by dozens of agricultural GMOs. This includes the crops which produce their pesticides and crops that survive while applying the specific herbicides (which is used to kill weeds).

There are studies done by the United Nations, the U.S. National Academy of Sciences, the European Union, the American Medical Association, U.S. regulatory agencies, and other organizations. In those studies, they found that GMO foods are safe but the skeptics contend that it is still too early to judge the long-term health and ecological effects of such crops.

2.5.3. Soil Science

There were studies done by the agronomists on sustainable ways to make soils much more productive. They tend to classify the soils and further reproduce them for determining whether they contain substances that are important for the plant growth like the compounds of N, P, and K and some other macro and micro-nutrients.

If any specified soil is deficient in such substances, then the fertilizers can be used to provide them. The soil science also includes the investigation of the movement of nutrients through the soil, the development of roots and their relation with the soil and the amount of the nutrients which are absorbed by the roots of plants.

Soil is said to be a mixture of different materials which include minerals, water, air, and rock. These lie on the top of the land which contains living and dead matter. Soil science can be defined as the study of the natural resources found on the surface of earth called soil. The soil science includes formation and mapping along with chemical, soil classification, biological, physical, and fertility properties of soils with regards to soil management.

There is a branch of soil science which includes chemistry, formation, morphology, and classification of soil and that is pedology. Whereas, there is edaphology which includes the influence of soil on microorganisms and plants. The nomenclature and the classification are done based on physical as well as chemical properties in the layers or horizons of soil.

In the soil classification system, the soil taxonomy uses color, texture, structure, and some other properties which are related to the soil surface. There are soil scientists who are qualified enough to evaluate the soil and further, they tend to interpret the soil in relation to the data for the purpose of understanding the soil resources in agricultural production.

Also to understand the environmental quality and management or even the protection of environment as well as human health. They have also conducted research in soil classification, tillage, drainage, and irrigation, plant nutrition, soil fertility and some other related areas.

Soil is considered to be a medium for plant growth, habitat for many different species, a filtration system for surface water and maintenance of atmospheric gases. Soil helps in sustaining life by providing food in the form of important minerals and nutrients; water and air which will help in the survival and growth of plants, worms, fungi, and bacteria.

It is important to study the soil resources for the environment, food, and fiber production. The techniques are required to be understood to improve soil conservation such as cover crops, crop rotation, planted windbreaks, and conservation tillage which affect both soil fertility as well as soil erosion.

Soil plays a key role in farms as it provides nutrients that are critical for the growth of agricultural as well as horticultural crops. Fertile soil is said to be very rich in nutrients and water. Thus, it is apt for agricultural practices. The fertile soil aids as the primary base of nutrients for healthy crops.

Rich soil usually contains pH along with some primary plant nutrients such as nitrogen, potassium, and phosphorus. It also comprises decaying content of organic matter together with some of the minor nutrients that help plants to grow better.

Some of the major functions that are associated with the soil are:

- Nutrient cycling;
- Water regulation;
- Ecological role in providing a medium for plant growth;
- Recycling organic wastes and nutrients;
- Modification of atmosphere;
- Water supply and purification;
- Habitat for soil organisms;
- Food production; and
- Flood control that eventually improves the quality of life.

Soil works as the natural protector of seeds and plants, dispersal, and germination of seeds in the soil ecosystem, the physical support system for the plants; retaining and transport of nutrients to the crops.

2.5.4. Soil Conservation

Additionally, there are certain methods that are developed by agronomists to preserve the soil. The methods can also be used to decrease the effects of erosion by water and wind.

Let's take an example. There is a technique which is known as contour plowing. This can be used to prevent soil erosion as well as conserve rainfall. The researchers in the field of agronomy look out for a number of different ways to use the soil in an effective manner for solving the issues or problems.

Such kind of problems consists of the disposal of human and animal wastes, water pollution and the build-up in the soil of pesticides. "Zero tillage" is a method that is now used to prevent erosion. Planting of grasses that are used to bind the soil, along with contours can be tried in steep slopes.

For a much better effect, contour drains of depths around 1 meter may aid in retaining the soil and prevent permanent wash off (Figure 2.4).

Figure 2.4. Soil conservation project in Bassin Zim (Central Plateau).

Source: Image by Flickr.

Soil conservation is a collection of techniques that are used to preserve the soil. Loss of soil and its fertility can be traced back too many causes such as erosion, over use, chemical contamination and salinization.

Unsustainable subsistence farming and the slash and burn clearing methods are used in some of the less developed regions. These methods can often cause loss of soil nutrients, deforestation, erosion on a large scale and at times, complete desertification.

Soil erosion removes the top layer of soil that is important for the organic matter, micro-organisms, nutrients, etc. that are required in the growth of plants. Conservation of soil is one of those steps that conserve the soil from being washed away.

The soil eventually ends up in aquatic resources bringing in fertilizers and pesticides used on the agricultural land. Healthy soil is critical for the growth of plants. Taking the required steps for the conservation of soil is a key part of an environmentally friendly lifestyle. There are different ways of soil conservation that can be done by the agricultural practices or the measures that people take at home.

2.5.4.1. Methods and Techniques of Soil Conservation

There are several different techniques that have been discovered throughout the years with the aim of preserving the nutrient level of the soil and preventing erosion.

- **Terrace Farming:** The terracing can be defined as a method of carving multiple, flat leveled areas into hills. There are many steps which are formed by the terraces which are surrounded by a mud wall to prevent run off as well as hold the soil nutrients in the beds. More commonly, they are found in lesser developed nations because of the difficulty while using mechanized farming equipment in the terraces. This method is said to be very popular in Asia for the purpose of planting rice.
- **Contour Plowing:** This is the technique that originated in ancient Phoenicia. The technique of Contour Plowing includes plowing grooves into the desired farmland. After that, the crop furrows are planted in the grooves which are further followed by the contours. It is said to be a very effective way for farmland on slopes to prevent run off and improve crop yields.
- **Perimeter Runoff Control:** This technique named as perimeter runoff control is said to be the practice of planting trees, shrubs, and ground cover around the perimeter of your farmland which impedes surface flows. Furthermore, this technique keeps nutrients in the farmed soil. There is a specialized way of handling the perimeter runoff and that is by using the grass way which uses surface friction for channeling and dissipating runoff.

- **Keyline Design:** The key-line design is a more enhanced version of Contour Plowing. In this technique, maximum water retention is achieved which is done by considering all the watershed properties while making the contour lines. The term 'Keyline' refers to the topographic feature that is connected to water flow. This, in turn, allows the water runoff to run directly into an existing water channel, and also, in preventing the soil erosion that is caused by the water.
- **Cover Crops/Crop Rotation:** There are certain cover crops namely turnips and radishes which are rotated long with cash crops to covering the soil all year. This technique round and produces green manure the replenishes nitrogen and some other critical nutrients. The weeds can also be suppressed and that can be done by using the cover crops.
- **Windbreaks:** Tall trees are planted in rows. This technique is used in dense patterns around the farmland as well as it helps in preventing wind erosion. The evergreen trees can provide year-round protection but there are deciduous trees also which can be adequate as long as foliage and so, is apparent during the seasons when the soil is bare.
- **Agrostological Measures:** In heavily eroded areas, the planting grass is referred to as agrostological measure. There are ley farming practices that cultivate grass in rotation along with regular crops to increase the nutrient level in the soils. When the grass is harvested, then it can be used as fodder for cattle. Considering heavily eroded soil, it is further recommended to grow grass for several years to let the soils naturally repair themselves.
- **Soil Conservation Farming:** A mixture of farming methods that intends mimics the biology of virgin land. These are the practices that can be used to prevent erosion as well as restore damaged soil and encourage plant growth. While eliminating the use of nitrogen, the fertilizer, as well as fungicides, can increase yields and also, protect the crops from drought and flooding.
- **Green Manures:** The green manures are considered as a few different crops which can be grown, not for produce or food usage, but for growing to fertilize the farmland on which it grows. This is the method that can improve the soil structure as well as suppress the growth of weeds.

- **No Till Farming:** The technique named as no till farming is said to be the method of growing crops around the whole year without changing the topography of the soil that can be by tilling or contouring. This is the technique that increases the amount of water that penetrates the soil and it can increase the organic matter of the soil. This, in turn, results in larger yields.
- **Salinity Management:** When the evaporation of water takes place from the soil, then it tends to leave behind its salt. This can result in damaging the soil and loss of nutrients. Using humic acids can prevent this or growing crops like saltbush can rejuvenate the soils and replace lost nutrients. Most of the time, the high levels of salt in the soil can be caused by changes that can be made to the water table by damming and other causes.
- **Stream Bank Protection:** During floods, most of the time, the stream banks can cave in. This can be prevented by constructing walls along the banks or the plant useful tree species which will prevent this in the upcoming future as well as prevent the soil loss down the stream.
- **Mineralization:** Sometimes, the crushed rock or chemical supplements are added to the farmland and this is done to help crops reach higher yields or full potential. Furthermore, this helps in combating mineral depletion. Normally this technique is used after flooding, and it brings substantial amounts of sediment. This sediment can damage the nutrient level of the soil.
- **Earthworms:** There are a number of great benefits which are provided by earthworms to the farm land. This is because the way they burrow under the ground and provide more are for water to rest after it has infiltrated the soil. When these worms excrete egesta, this sits in the soils and gives the crops many nutrients that are absorbed with the help of the roots of the plants. The earthworm casts contain a large amount of nutrients in comparison to any natural soil in the world. This can be the reason that it is important to invite them into the soils of farmland because they can eventually help in preventing erosion and then, it will lead to larger crop yields.
- **Reduction of Impervious Surfaces:** The driveways patios as well as paved pathways allow precipitation to flow freely off them. As the water flows, it picks up momentum and then, erodes any soil

in which it flows over after leaving the impervious surfaces. This in turn reduces the amount of these around your farmland which will prevent erosion.

- **Korean Natural Farming:** This method named as Korean natural farming takes advantage of natural as well as indigenous microorganisms for producing the fertile soils which yield high output and get rid of the need to use herbicides or pesticides. This method is used in the respective areas because of the improvement in soil health and output.
- **Dry Farming:** There are areas that have a very low amount of rainfall. In those areas, it is important to grow such crops that require very little water. This will lead to the preservation of the natural levels of moisture as well as nutrients in the soil.
- **Re-Establish Forest Cover:** When there are a large number of trees in a forest then it results in a vast network of deep roots and this, in turn, offers a long term solution for soil erosion. One more benefit is the windbreak that can be provided from these trees.
- **Rain Gardens:** A rain garden is said to be a shallow depression in the land which holds and collects the water running from impervious surfaces and it prevents erosion while saving the nutrients that inevitably can get washed away. Also, this gives a good bed for growing wetland plants.
- **Maintaining PH Levels of Soil:** The contamination of soils can take place because of the acid rains and some other pollutants which can result in loss of soil fertility. It is important to use a pH indicator monthly to check the levels of acids in the soil as well as the soils should be treated with eco-friendly chemicals for preventing a loss of crops and low yields.
- **Prevent Overgrazing:** It is important to take care that tries not to let overgrazing happen by moving herds around often. When overgrazing occurs, then the plant gets hardier and more nutritious species of forage to rebuild the soil. These crops can be harvested and further, feed them to the grazers during the winter season.
- **The Sharing of Knowledge:** The more developed countries can share their farming knowledge which they have gained throughout the years and they should do that with the lesser developed nations of the world. This will result in a better quality of soil worldwide

and can help to prevent famine and also, solves the food crisis in some of the areas of the world.

- **Indigenous Crops:** The growth of indigenous crops is considered to be a good way while conserving the soil, because the plants have a natural need for the nutrients in the soil in that area, they help in preventing soil erosion. If non-indigenous crops are grown, then there is a recommended tip that indigenous crops should be planted around them, to prevent soil erosion.

2.5.5. Agroecology

Agroecology is considered to be a scientific discipline that uses ecological theory to study, design, manage, and evaluate agricultural systems which are said to be productive and resource conserving.

The agroecological research considers the interactions of all the important biophysical, technical, as well as socioeconomic components of farming systems and regards these systems as the fundamental units of study, where mineral cycles, energy transformations, biological processes, and sociological relationships are analyzed as a whole in an interdisciplinary fashion.

Agroecology is majorly related to work in the areas of sustainable agriculture, alternative food systems, organic farming, and the development of alternative cropping systems.

Agroecology tends to share much which is in common with some other approaches to sustainable farming. The agroecology is said to be farming that "centers on food production that makes the best use of nature's goods and services while not damaging these resources."

When the farming works with the local systems then it thrives. For example, improvement in soil and plant quality by using the available biomass and biodiversity, instead of battling nature with chemical inputs. Agroecological farmers try to improve food yields for balanced nutrition, strengthen fair markets for their produce, furthermore, enhance healthy ecosystems, and also, build on ancestral knowledge and customs.

The promoters of agroecology attempt to nurture a healthy landscape in which the food and fiber of the world are grown. They are further guided by an ethos of the bio as well as cultural diversity which in turn features small farmer-centered applied research and also, policies that protect their livelihoods.

There are worldwide, scientists, NGOs, grassroots organizations, consumers, universities, and public agencies that are working with farmers for constructing sustainable as well as nutritious food systems that are based in agroecology.

Now, there are unprecedented opportunities for the advancement of agroecology globally. Most of the time, the corporate food system has negative impacts on the health of people, on the environment and the well-being of family farmers.

Furthermore, agroecology is considered to be both mitigation and adaptation strategy for climate change. The healthier food is continuously in demand by the consumers and they tend to require a closer connection to food producers. There are social movements that are taking place around the globe and many of them have significant leadership by women's and indigenous organizations which are coalescing in campaigns for a healthy food system built on an environmental as well as human rights ethos. The demand for agroecology is increasing.

Fusion of Agronomy and Ecology: After the period of 1970s, the agronomists saw the value of ecology. The ecologists started to use the agricultural systems to study plots, the studies in agroecology grew more rapidly. Gliessman describes that the innovative work of Prof. Efraim Hernandez X. He was the professor who developed research based on indigenous systems of knowledge in Mexico, which led to education programs in the agroecology.

In the year 1977, Prof. Efraim Hernandez X. explained that the modern agricultural systems had lost their ecological foundation when the socio-economic factors were considered to be the only driving force in the food system.

The acknowledgment that the socio-economic interactions are considered to be one of the fundamental components of any agroecosystems and this came to light in the year 1982, along with the article Agroecologia del Tropico Americano by Montaldo. Further, the author argues that the socio-economic context cannot be separated from the agricultural systems while designing agricultural practices.

In the year 1995, in Sustainable Agriculture and Integrated Farming Systems, Edens solidified this idea further proving his point which he did by devoting special sections to the economics of the systems, ecological impacts and ethics and values in agriculture.

At the end of the year 1985, the year was considered to be a fertile as well as creative year for the new discipline. For example, in the same year, Miguel Altieri integrated the way in which consolidation of the farms and cropping systems impact the pest populations.

Additionally, Gliessman emphasized that the socio-economic, technological, and ecological components give rise to the choices of producer related to food production systems. These were pioneering agroecologists who have helped in framing the foundation of what we today consider the interdisciplinary field of agroecology.

2.6. GIS IN AGRONOMY

Agronomy is a branch of agriculture which is known to deal with soil and crops. Soil differs in space and hence, it is highly responsible for variation in the yield as well as the growth of crops on the agricultural lands. Such kind of variation in the production of crops planted and observed in the same area of land under the optimized environmental conditions has a major concern to the farmers.

Spatial variations in the status of soil nutrients as caused by topography, texture, and management practices of soil have been observed all across the fields. Therefore, the need to segregate the field into site specific management units by using Geographical Information Systems (GIS) for effective soil and crop management to obtain optimum productivity (Figure 2.5).

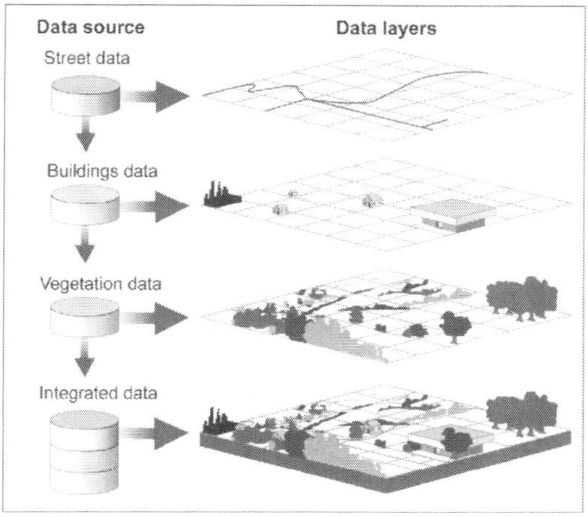

Figure 2.5. Visual representation of themes in GIS.

Source: Image by Flickr.

Over the past few years, farming directions, size of fields, location of fences, fertility programs and rotation have modified the nutritional status of the farms. As a result, the productivity of the soil has also been affected.

Despite all these factors, traditional agricultural activities treat the entire field in a uniform manner in context to the use of fertilizers, soil amendments, pesticides, and various other chemical products.

The application of GIS will aid the farmers to get rid of over or under use of fertilizers and several other agrochemicals uses. In the field of agronomy, the potential of GIS is quite large. Although, the community of GIS users in the field of agronomy is quite small as compared to some other sectors.

2.6.1. Importance of GIS to Agronomy

GIS could play a key role in the field of agronomy at different levels because of the fact that it can be used to study the status of nutrients of particular fields so as to arrive at exact requirements for the external use of nutrients.

Agronomic activities are said to be spatial and the requirement of placing site-specific information in a long-term and spatial perspective would need special models that can be further utilized to measure spatial variation in growth of crops and observe variations in trend with a time scale suitable for guiding decisions.

According to Haddish Melakeberhan, the application of GIS in agronomic practices aids in managing the information intensive environment in the production of crops by integrating site-specific management with the computer software modeling especially for analyses and interpretation of different inputs and outputs.

As opposed to the typical manual adjustment of the farmers, GIS helps them in managing with-in field variable rate application, which results from spatial variation in crop yields within a field. Thus, GIS improves the assessment as well as understanding of variations in a field crop.

According to Bullock DS, Lowenberg-DeBoer and Swinton (2002), GIS can be applied for assembling various layers of the information like soil nutrients, moisture content elevation and topography to produce a map to show what kind of factors impact the yield of crops.

In the year 2002, it was also reported that the production can be estimated for future reference. The economic inputs and outputs can be measured depending upon the expected production. This will eventually have a great potential for saving the costs spent on over used fertilizers that otherwise could have been used in some other field.

2.6.2. Applications of GIS in Agronomy

According to Hartkamp, White, and Hoogenboom, (1999), GIS applications have primarily developed from hydrological applications during mid-1980s to the current diverse variety of applications in the field of agronomy as well as natural resource management research.

Certain examples of GIS applications include:

- Atmospheric modeling;
- Climate change;
- Sensitivity and/or variability studies;
- Characterization and zonation;
- Hydrology;
- Water quality;
- Water pollution;
- Soil science;
- Spatial yield calculation—regional, global; and
- Precision farming (spatial yield calculation).

A great number of studies have been reported on the GIS application on the cultivation practices of different varieties of crops.

In the year 2010, the authors reported that the application of GIS to the management of fertility of soil planted to tea where digitized maps of the soil pH, phosphorus, potassium, and organic matter were made by using the Arc MAP software.

According to Senthurpandian, Jayaganesh, Srinivas, Palani, and Muraleedharan, (2010), it would be great for tea growers in certain locations for measuring the requirements of fertilizers. In the year 2010, it was reported that certain measures may be needed to reduce to the desired level of the pH of fields having pH > 5.5. In the year 2013, a geodatabase was developed using GIS mapping.

This was mainly done to provide the soil quality monitoring depending upon the data of agrochemical soil survey to observe land cover/soil quality changes between periods of soil survey.

In the work of Kokhan, Moskalenko, and Shkvir, GIS mapping for soil quality evaluation. 2nd International Conference of Informatics and Management Sciences (2013), ArcGIS was used for mapping the quality of soil. It was reported that soil data can be easily handled as well as analyzed by using ArcGIS as they are spatial.

It was also reported that there were no critical changes in the humus easily hydrolysable nitrogen content within the period between the last two soil agrochemical surveys. While in the year 2012, a GIS-based decision support system was applied for establishing the potentials as well as limitations of different soils for production of crops. In 2009, GIS was used to control soil erosion where certain factors and elements affecting erosion were studied by examining the numerical maps of different parts of a basin.

2.7. CONCLUSION

Agronomy is considered to be the science and technology which is related to the production and usage of plants in agriculture for food, fuel, and land restoration. It is both a humanitarian career as well as a scientific one.

The plant genetics, plant physiology, soil science, and meteorology are some of the areas in which the agronomy has encompassed work.

It is said to be the application of a combination of sciences such as biology, economics, chemistry, earth sciences, ecology, and genetics. In today's world, the agronomists are involved in a number of issues which include producing food, managing the environmental impact of agriculture, creating healthier food, extracting energy from plants and distribution of agriculture.

Most of the time, the agronomists are specialized in areas such as irrigation and drainage, plant physiology, plant breeding, soil fertility, soil classification, insect, and pest control and weed control.

The conclusion that can be drawn from the chapter is that in today's scenario, agronomy is said to be important because the application of technology in agriculture has made things easy for the farmers. There are certain principles of agronomy and sub-discipline of agronomy which have all been discussed in the chapter.

REFERENCES

1. Agric.ucdavis.edu. (n.d.). About Agronomy. [online] Available at: http://agric.ucdavis.edu/About_Agronomy/(accessed on 2 April 2020).
2. AgroEcology Fund, (n.d.). What is Agroecology?—AgroEcology Fund. [online] Available at: https://www.agroecologyfund.org/what-is-agroecology (accessed on 2 April 2020).
3. Allard, R., (1999). Plant Breeding | History, Applications, and Methods. [online] Encyclopedia Britannica. Available at: https://www.britannica.com/science/plant-breeding (accessed on 2 April 2020).
4. Bareja, B. G., (2011). What is Agronomy: Definition and Comparison with Horticulture. [online] CropsReview. Available at: https://www.cropsreview.com/what-is-agronomy.html (accessed on 2 April 2020).
5. Chandrasekaran, B., Annadurai, K., & Somasundaram, E., (2010). A Textbook of Agronomy. [eBook] p. 1. Available at: https://www.dphu.org/uploads/attachements/books/books_2274_0.pdf (accessed on 2 April 2020).
6. Conserve Energy Future, (n.d.). What is Soil Conservation and Methods of Soil Conservation-Conserve Energy Future. [online] Available at: https://www.conserve-energy-future.com/methods-of-soil-conservation.php (accessed on 2 April 2020).
7. Cststudy.blogspot.com. (2018). Agronomy: Scope and Principle. [online] Available at: http://cststudy.blogspot.com/2018/02/agronomy-scope-and-principle.html (accessed on 2 April 2020).
8. Ecoursesonline.iasri.res.in. (n.d.). AE: Lesson 18. Definition and Scope of Agronomy. [online] Available at: http://ecoursesonline.iasri.res.in/mod/page/view.php?id=1542 (accessed on 2 April 2020).
9. Encyclopedia Britannica, (1998). Biotechnology | Definition, Examples, and Applications. [online] Available at: https://www.britannica.com/technology/biotechnology (accessed on 2 April 2020).
10. JLI Blog, (2019). What is the Importance of Soil Science in Agribusiness? | JLI Blog. [online] Available at: https://www.jliedu.com/blog/soil-science-agribusiness/(accessed on 2 April 2020).
11. John Deere MachineFinder, (2013). The Science of Agronomy: How Agriculture Responds to the World Around Us. [online] Available at: https://blog.machinefinder.com/11023/the-science-of-agronomy-how-agriculture-responds-to-the-world-around-us (accessed on 2 April 2020).

12. Oshunsanya, S. O., & Oghene, A. O., (2015). GIS Applications in Agronomy. [online] IntechOpen. Available at: https://www.intechopen.com/books/geospatial-technology-environmental-and-social-applications/gis-applications-in-agronomy (accessed on 2 April 2020).
13. Rana, S., & Kumar, S., (2014). Research Techniques in Agronomy. [eBook] p. 2. Available at: http://hillagric.ac.in/edu/coa/agronomy/lect/agron-592/592-Research-Techniques-in-Agronomy-SSR-SKG.pdf (accessed on 2 April 2020).

CHAPTER 3

Agricultural Meteorology

CONTENTS

This chapter named as 'Agricultural Meteorology' basically includes all the basics related to meteorology. In the first half of the chapter, it mentions the scales, history, and scope of meteorology. In the second half of the chapter, it includes the benefits that can be gained from the agro-meteorology and also the meteorological parameters that are used in agriculture.

The readers will get insights about the meaning and definition of agricultural meteorology, the scope of agricultural meteorology, the importance of agricultural meteorology and also the applications of agricultural meteorology.

3.1. INTRODUCTION

In agriculture, the application of meteorology is important. The reason is that every facet of agricultural activity tends to depend on the weather. Some other agricultural decision-maker gets to benefit from the agrometeorological applications. The weather affects everyone to some extent of degree. Unsatisfactory weather conditions frequently hamper the plans of the farmer, the market gardener, the construction engineer, and the lumberjack. Hence, everyone has tried to weather control (Jaybhaye et al.2012-Pralhad Jaybhaye, Pankaj Raunadl and Ravi Andhale. 2012. Handbook of Agrometeorology. Continental publication, Pune, pp)

This includes government policymakers that seek to make sure that the relevant food supplies, food prices are affordable for consumers and enough farm income for the farmers, and, to decrease the effect of the agricultural practices on the environment.

In international organizations, the decision makers also use the applications of the meteorological to make sure about the food security and also, to react to the potential famine situations. At a certain level, there are certain objectives that can be achieved only through the active cooperation among the National Meteorological and Hydrological Services (NMHSs), farmers, and their associations, agricultural extension services, agricultural research institutes, universities, and industry.

When research is being done in agrometeorology then there arises a central problem and that is how to convert the knowledge of the physical environment into reliable means to make decision with respect to crop management.

Considering the recent time, it can be said that the management continues to be a much important aspect while developing efficient agricultural production systems in a tropical country like India is. The countries where the uncertainty of weather is much more pronounced.

The productivity of crop is said to be a function of several input variables. Among these variables, the weather is the most important one. Information related to the weather condition is likely to prevail and that too during the time of the growing season.

This is useful for taking any decisions in agricultural planning as well as management. There is rainfall which tends to contribute to the soil moisture and it plays an important role in rainfed farming.

The agricultural productivity depends on rain in a substantial manner and also its distribution pattern. In the eastern plateau region, the climate is humid, sub-tropical. This is characterized by an intense hot as well as dry summer along with the large winter water deficits. A great variation can be seen in the 900 to 1500 mm. Considering the plateau region, the average annual rainfall is equal to 1372 mm.

In any of the given region, it is important to collect the scrutinize and analyze the historical weather data which is available in the region to identify:

- The length of the possible cropping period when availability of the water is not a severe constraint, and
- when water is considered to be the limiting factor for crop production under rainfed conditions.

Under the regions, which are sub-humid and sub-tropical, the crops are grown under the rainfed conditions. Therefore, the length of the cropping season which is available for crop production depends on:

- Distribution of the rainfall during the rainy season;
- Nature of soil with water holding capacity;
- End of the rainy season and post monsoon rainfall pattern;
- Onset of rainy season; and
- Evaporative demand of the atmosphere (Ramana Rao, 1988).

It is important to have knowledge of probable dates of commencement and end of rainy season as well as the duration of intermittent dry and wet spells. This knowledge can prove to be very much useful while planning a number of agronomic practices.

Such kind of information can prove to be useful to minimize the risks to crops as well as optimum utilization of limited resources such as labor and fertilizer having suitable cropping systems in different topo sequences.

In the life span, there are some important periods of each crop that is from sowing to harvesting. This is considered to be the most important for the paddy, the dominant crop in the plateau area. Therefore, if one has the knowledge of pre-sowing and sowing rains, it's quantum as well as distribution pattern during the phases of critical growth having the frequency of the dry spell occurrences, then they can make an effective crop/varietal choice with less risk of failure in monsoon (Kharif) season.

The knowledge of rainfall characteristics that occur after monsoon, along with the soil moisture retention pattern in a number of different topo sequences, can guide people so that they can adopt the effective sequential cropping systems.

Let's take an example of the Bihar plateau. This type of knowledge becomes all the more important in these areas. In such areas, in spite of annual rainfall of 1200–1300 mm, success of paddy (on the basis of cropping systems) majorly depends on the rainfall because of the non-existence of water harvesting facilities.

The weather parameters significantly influenced the crop growth and development and finally reflected into significant variation in the Bt cotton seed yield. **The** prolonged dry spell may occur initially or at mid-season or terminally which generates moderate to severe drought condition and which coincides at least one critical physiological stage of cotton. It causes distinguished reduction in seed yield of rainfed Bt cotton. Koraboina Mamatha* and P. R. Jaybhaye Impact of Drought Weather Condition on Bt Cotton Growth, Development and Yield Int.J.Curr.Microbiol.App.Sci (2018) Special Issue-6: 2332-2338

3.2. METEOROLOGY

Meteorology can be defined as the study of the atmosphere, atmospheric phenomena, and the atmospheric effects on the weather. The atmosphere can be referred to as the gaseous layer of the physical environment which surrounds a planet.

The atmosphere of the Earth is roughly 100 to 125 kilometers (which is equal to 65–75 miles) thick. The atmosphere can expand too much farther but there is gravity which keeps it from expanding much farther.

Meteorology is said to be a sub-discipline of the atmospheric sciences. The term 'atmospheric sciences' covers all the studies related to the atmosphere.

A sub-discipline can be defined as a specialized field of study which is within a broader subject or discipline. Climatology and aeronomy are also considered as the sub-disciplines of the atmospheric sciences.

The climatology mainly focuses on the way in which the atmospheric changes define and alter the climates of the world. The study of the upper parts of the atmosphere, where unique chemical as well as physical processes occur, can be referred to as aeronomy.

The study of meteorology majorly focuses on the lower parts of the atmosphere (primarily the troposphere). In the lower parts of the atmosphere is the area where most of the weather takes place (Figure 3.1).

Figure 3.1. Meteorology studies thunderstorm.

Source: Image by Pixabay

There are scientists referred to as meteorologists who use scientific principles to observe, explain, and further forecast the weather. Most of the time, they focus on atmospheric research or operational weather forecasting.

There are several sub-disciplines which is covered by the research meteorologists. This includes climate modeling, air quality, remote sensing, climate change, and atmospheric physics. Also, the research on the relationship possessed by the atmosphere and the climates, oceans, and biological life on Earth.

The research is then used by the forecasters. They also use the atmospheric data to assess the current state scientifically of the atmosphere and further, make predictions of its future state.

Both at the surface of earth and above, the atmospheric conditions are measures and for that, a number of sources are used named weather stations, buoys, ships, radar, aircraft, weather balloons and satellites. The data thus received is transmitted to centers all over the globe which produce the computer analyses of global weather.

The analyses are thus passed further to the national as well as regional weather centers. These centers feed this data into computers that model the future state of the atmosphere. This transmission done for the information represents the way in which the weather and the study of it takes place in a number of interconnected ways.

3.3. SCALES OF METEOROLOGY

There are a number of different scales of space and time in which the weather occurs. There are four meteorological scales which include microscale, mesoscale, synoptic scale, and global scale. Most of the time, the meteorologists focus on a particular scale in their work.

3.3.1. Microscale Meteorology

The microscale meteorology mainly focuses on the phenomena which range in size from a few centimeters to a few kilometers and that have a short life span (that can be less than a day). Very small geographic areas, and the temperatures and terrains of those areas, are affected by this phenomenon.

Most of the time, the microscale meteorologists study the processes which occur in between soil, vegetation, and the surface water near ground level. They further measure the transfer of heat, gas, and liquid that are present between these surfaces. Often, microscale meteorology includes the study of chemistry also.

There can be one example of the microscale meteorology and that is tracking the air pollutants. MIRAGE-Mexico is said to be the collaboration between the meteorologists of the United States and Mexico.

This is a program that tends to study the chemical as well as physical transformations of gases and aerosols in the pollution that is present in the

Mexico City. MIRAGE-Mexico tends to use the observations from ground stations, satellites, and aircraft to track the pollutants.

It may explain though the example of the agrometeorological zones of India, which has different weather conditions as per season as well as zone. Within a state there are may be many microclimatic conditions or zones, which are suitable to grow the specific crop commercially.

The soil, water and weather decides the productivity and test of agricultural products; and also agricultural processed products. Hence, Umarkhyyam Saympian was won the gold medals in different wine fairs in Europe. Likewise, Basmati of Bhandara Dist. (Vidharbha), Ambebahar of Junner (Pune) of rice cultivar; Jawar of Solapur,[Marathawada (i.e. Maldandi (M-35)] and Jawar cultivar and Jaggary of Solapur give excellent test which could not found anywhere in the country. In fruit crops Alphanso of Devgad (Konkan) and Kesar of Marathwada, sappota of Dahanu (Thane), fig of Sasawad (Pune), Santra of Nagpur gives excellent test and quality fruits. All these examples indicate that the variation of the test and quality of yield and processed products of agricultural produce is due the variation in 'Agroclimatic zone'. (Jaybhaye et al.2012-Pralhad Jaybhaye, Pankaj Raunadl and Ravi Andhale. 2012. Handbook of Agrometeorology. Continental publication, pune,pp)

3.3.2. Mesoscale Meteorology

The mesoscale phenomena range in size from a few kilometers to around 1,000 kilometers (that is 620 miles). There are two important phenomena namely mesoscale convective complexes (MCC) and mesoscale convective systems (that is MCS). Both of these phenomena are caused by convection. The convection is said to be an important meteorological principle.

Convection is referred to as a process of circulation. The fluids which are warmer and less dense tend to rise and the fluids which are colder and denser tend to sink. The fluid that is most studied by the meteorologist is air. (Here, any substance that flows is considered to be fluid). There is a transfer of energy, heat, and moisture that is because of convection which is the basic building blocks of weather.

In both that is MCC as well as MCS, there is a large area of air and moisture which is warmed during the middle of the day, at the time when the angle of the sun is at its highest. Along with the rise in the warm air into the colder atmosphere, it tends to condense into clouds, further turning water vapor into precipitation.

An MCC is said to be a single system of clouds that can reach the size of the state of Ohio and it produces heavy rainfall and flooding. An MCS can be defined as a smaller cluster of thunderstorms which lasts for long hours. Both tend to react to the unique transfers of energy, heat, and moisture which are caused by convection.

There is a program named as the Deep Convective Clouds and Chemistry (that is DC3). This is a field campaign that will study the storms and thunderclouds in Colorado, Alabama, and Oklahoma. This is a project which considers the way in which convection influences the formation as well as the movement of storms. This includes the development of lightning. Also, it will study the impact on aircraft and flight patterns. The DC3 program tends to use the data which is collected from the research aircraft which is able to fly over the tops of storms.

3.3.3. Synoptic Scale Meteorology

The synoptic scale covers almost several hundreds or even thousands of kilometers. The synoptic scale is used in the high and low-pressure systems seen on local weather forecasts. The pressure is much similar to convection which is an important meteorological principle. The roots of large-scale weather systems are as diverse as the hurricanes and bitter cold outbreaks.

When the atmospheric pressure at the surface of the Earth is less than its surrounding environment, the low-pressure system occurs. The area where there is high pressure seeks low-pressure systems with winds and moisture. The Coriolis force and friction along with this movement make the system to rotate counter-clockwise in the Northern Hemisphere and clockwise in the Southern Hemisphere, creating a cyclone.

This movement, in conjunction with the Coriolis force and friction, causes the system to rotate counter-clockwise in the Northern Hemisphere and clockwise in the Southern Hemisphere, creating a cyclone. Cyclones have a tendency to rise upwards in a vertical motion. This eventually forms clouds as it rises upwards, expands, and condense into water vapor. The majority of the weather events are caused due to this type of event.

Low pressure system causes hurricanes that develop over the tropical water in the Western Hemisphere. The moisture present in the sea is sucked up by the system. This causes convection to take place. This, in turn, causes wind speeds to increase and pressure to fall. When the speed of the wind reaches over 119 km per hour, then it is known as a hurricane.

In Western Hemisphere, one of the most devastating natural disasters is hurricanes. There is a National Hurricane Center in Miami that regularly issues forecasts and reports on all tropical weather systems. During the season of hurricane, the specialists of a hurricane issues the reports and forecasts for the tropical storm in the western tropical Atlantic and eastern tropical Pacific. Most of the business officials, government officials from the United States, the Caribbean, Central America, and South America rely on forecasts from the National Hurricane Center.

The high-pressure systems occur when the atmospheric pressure at the surface of the Earth is greater than its surrounding environment. This pressure lets the hurricane to move in downward vertical motion that allows it for dry air and clear sky.

The high pressure system leads to the extreme cold temperature that develops over the Arctic and the northern hemisphere. Since the Arctic is over the ice and snow covered ground, the air blown there is very cold. This cold air pushes against the surface of the earth with extreme pressure that prevents any moisture from air heat from staying within the system.

They have been mini research by the meteorologist regarding semi-permanent areas of high pressure. For instance, the Azores high is considered as a stable region with high pressure around the Azores. It is situated in the mid Atlantic Ocean. The arid temperatures of the mediterranean basin and the heat waves of summer in Western Europe are due to the Azores high.

In concern to the Synoptic Scale, the monsoon climate is one of the examples. Along with the Indian continental many countries or a part of country are included in monsoon climate. Weather and climate are the most important input for the agriculture production and agricultural enterprises under rainfed farming system are mainly dependent on monsoon rainfall and the vagaries of monsoon however, keep the profitability always at stake. P. R. Jaybhaye, S. D. Gawali, V. K. Mohite, P. B. Shinde and B. V. Asewar. 2016. Impact of Changing Rainfall Trend on Yield of Mango and Sweet Orange in Marathwada Region. Indian Journal of Ecology (2016) 43 (Special Issue-1): 250-253

3.3.4. Global Scale Meteorology

The weather patterns which are related to the transport of heat wind and moisture from the tropics to the poles is the global scale phenomena. In this, there is an important pattern which is known as a global atmospheric circulation. The global atmospheric circulation is the movement of air in a large scale that helps distributes thermal energy across the surface of the earth.

Hadley cells are the main reason behind this atmospheric circulation. The tropical and equatorial convection pattern is the Hadley cells. The cooler is pushed down words while the convection drives the warm air in the atmosphere in a constant loop each loop is called a Hadley cell.

The flow of trade winds is determined by Hadley cells which the meteorologists forecast. The strength of the trade winds is paid very close attention by the businesses which are into exporting products across oceans as it will help the ship travel faster. There is a term called westerlies which are the winds that are blown from the west in the midlatitudes. The trade winds blow from the North East and the South East closer to the equator.

The long-term climate patterns that disrupt global atmospheric circulation is studied by meteorologists. El nino pattern was discovered by meteorologists. The ocean currents and trade winds across the Pacific Ocean constitutes El nino. Now it occurs roughly after every 5 years that disrupts global atmospheric circulation and affects the local weather and economies from Australia to Peru.

El nino is linked with the changes in the air pressure in the Pacific Ocean which is known as southern oscillation. Near to the coast of America, the pressure of airdrops over the Eastern Pacific while on the other hand the air pressure rises over the Western Pacific near to the coast of Australia and Indonesia. This weakens the trade winds.

There is extreme rainfall in the Pacific Nations. The warm ocean currents reduce the number of fishes in the sea which is dependent on the nutrient rich upwelling of cold water to thrive. There has been an experience of drought and devastating agricultural production in the Western Pacific Nations.

The better understanding of the material logical processes of El-Nino eventually helps farmers fishes and coastal residents in preparing the climate pattern.

3.4. HISTORY OF METEOROLOGY

The development of meteorology is majorly connected with the developments in science, math, and technology. The first major study of the atmosphere was written by the Greek Philosopher which was done around 340 BCE.

There were a number of ideas given by Aristotle which were incorrect. The reason was that he did not believe it was important to make scientific observations.

In the scientific method, there was a growing belief and that profoundly changes the study of meteorology 17th and 18th centuries. Evangelista Torricelli was an Italian Physicist. He observed that the changes in the air pressure were connected to the changes in the weather. In the year 1643, Torricelli invented the barometer.

Further, he used it to measure the pressure of air. Still, the barometer is considered as the key instrument while understanding as well as forecasting the weather systems. There was a German Physicist namely Daniel Fahrenheit. In the year 1714, he developed the mercury thermometer. Such kind of instruments made accurate measurement possible for two atmospheric variables.

There was no such way that can quickly transfer the weather data until unless the telegraph was invented. The telegraph was invented by Samuel Morse which was an American inventor in the mid-1800s.

This was a new technology that was used and using this, the meteorological offices were able to share the information as well as they were able to produce the first modern weather maps. Such maps were combined and also, they displayed much more complex sets of information such as isobars (that is the lines of equal air pressure) and isotherms (that is the lines of equal temperature).

Having such large-scale weather maps, the meteorologists were able to examine a broader geographic picture of the weather and also, make much more accurate forecasts.

In the period of 1920s, there was a group of Norwegian meteorologists. They developed the concepts of air masses and fronts that are considered to be the building blocks of modern weather forecasting. The basic laws of physics were used by these meteorologists while discovering that the huge cold and warm air masses move and tend to meet in patterns. These patterns are the root of a number of weather systems.

During the period of World War I and World War II, the military operations brought a number of great advances to the meteorology. These operations were successful and their success was highly dependent on the weather over vast regions of the globe. There were heavy investments done by the military in the training, research and new technologies to improve their understanding of weather.

Among these new technologies, the most important was the radar. This was developed to detect the presence, direction, and speed of aircraft and ships. Radar was used since the end of World War II and this improved the detection of the presence, direction, and speed of precipitation and wind patterns.

The technological developments which took place in the period of 1950s and 1960s made it easier as well as faster for the meteorologists to observe and predict the weather systems on a large scale.

During the period of 1950s, the creation of first models of atmospheric conditions took place using the computers by running hundreds of data points through the complex equations. Such kind of models was able to predict the weather on large-scale, such as the series of high- and low-pressure systems that surrounds our planet.

The first meteorological satellite is named as TIROS I. this satellite provided the first accurate weather forecast from space in the year 1962. Furthermore, the success of TIROS I prompted the creation of much more sophisticated satellites.

They have the ability to collect well transmit data having extreme accuracy and speed. This has made them indispensable to meteorologists. There are advanced satellites as well as the computers which process their data. These are primary tools that are used in meteorology today.

Meteorology is considered to be the study of atmosphere of the Earth and also, about the variations in temperature and moisture patterns which in turn produce various weather conditions. There are some major subjects who study the phenomena of precipitation (rain and snow), thunderstorms, tornadoes, and hurricanes and typhoons.

There are a number of different ways in which the importance of meteorological events can be felt. Let's take an example, a drought happens, and it leads to a shortage in water, crop damage, low-river flow rates and also increase the wildlife potential.

Along with this, such kinds of effects can further result in restricted river travel, saltwater infiltration in aquifers and coastal bays, stress on a number of different plant and animal species, economic hardship, population shifts and also, political unrest.

On human activity, the important impact of weather has resulted in the development of the not so certain science of weather forecasting.

The word 'meteorology' is derived from the Greek word 'meteoron.' This word refers to any phenomenon that occurs in the sky. All of phenomenon which occurred above the ground is mentioned in the Aristotle's Meteorologica (340 BC). The Astronomy included the study of meteors or shooting stars and later on, this study became a separate discipline.

The science of meteorology was further limited ultimately to the study of atmosphere. Different weather phenomena are still referred to as 'meteors' and these include hydrometers (which include liquid or frozen water-rain, snow, and snowflakes, clouds, fog), litho meteors (which include dry particles such as sand, dust or smoke), photo meteors (which includes optical phenomena such as mirages, halos, rainbow, coronas), and electro meteors (which include electrical phenomena such as lightning, Saint Elmo's fire).

The modern meteorology highlights the typical weather patterns which are being observed. These patterns include thunderstorms, extratropical cyclones, hurricanes, fronts, typhoons, and different tropical water waves. Usually, the meteorology is said to define as well as study the physical foundation for individual events.

Contrastingly, the climatology is said to define as well as study the origin of atmospheric patterns which is observed over the period. There are a number of important phenomena such as monsoon, which are considered in both the meteorology as well as climatology. The reason for this is that they exhibit the large changes in seasonal time scales.

3.4.1. Scope

A number of fields of science and engineering were drawn by the effort to understand the atmosphere and its processes. The dynamic meteorology can be defined as the study of atmospheric motions. This study uses the equation while describing the behavior of a compressible fluid (air) on a rotating sphere that is Earth.

In this study, there is one important complication and that is the fact that the water in the atmosphere changes back and forth between solid, liquid, and gas in a much complicated manner. Such kind of changes tends to greatly modify the equations which are used in dynamic meteorology.

Physical meteorology which is often named as atmospheric physics tends to deal with a number of areas of study which are specialized in a particular field. For example, the study of clouds and the study of its different forms of hydrometeors include investigations into the behavior of water in the atmosphere.

There is a study that is of radioactive transfer. This study is concerned with the fundamental source of energy which drives atmospheric processes such as solar radiation. It also drives the ways in which radiant energy is used and dissipated in the atmosphere. Some other specialized disciplines tend to deal with the phenomena which include light (atmospheric optics) and sound (atmospheric acoustics).

There are some branches of meteorology which are defined in terms of size of the phenomena which are being studied. Lets' take an example. The micrometeorology is majorly concerned with the study of the small-scale interactions between the lowest level of atmosphere and the surfaces with which it comes in contact.

The mesoscale meteorology is the study that deals with the phenomena of immediate size, for example, thunderstorms, and mountain winds. The synoptic meteorology is the study which deals with the larger processes like the high- and low-pressure systems and their fonts and many more up to the study of entire atmospheric circulation for the time scales of a few days.

The weather forecasting is said to be the predictive aspect of meteorology and it derives from these disciplines.

There are some other branches also which deals with the phenomena in any particular locations like the equatorial areas, maritime regions, the tropics, the poles and mountains and the coastal areas. Also, the upper atmosphere is studied separately.

There are various disciplines which are focused on taking observations using some specified technologies such as radio, radar, and artificial satellite. There is one computer technology that is used extensively and its uses include numerical weather prediction, display systems, and interactive data analysis.

The chemical behavior of the atmosphere is studied in the atmospheric chemistry and this has gained much importance because of the inadvertent changes which caused by the humans in the molecular composition of the atmosphere.

There are situations that have gone beyond the status of local problems which include changes in ozone (and the ozone layer) and carbon dioxide concentrations and also, increased levels of acid rain and these have become regional or global issues.

There are meteorological studies that are carried out along with a number of different fields that are related to the environment. Such studies include aeronautics, architecture, agriculture, ecology, ballistics, forestry, energy production, hydrology, medicine, and oceanography.

A number of these related fields are required to determine the effects of weather at a particular time and place. However, there are some other studies also, which include hydrology and oceanography which affect the meteorological events and this is done by modification of the atmospheric conditions at the surface of Earth.

3.5. AGRICULTURAL METEOROLOGY (FIGURE 3.2)

Difference: Agrometeorology and Meteorology

Meteorology	Agrometeorology
Branch of Atmospheric physics	Branch of agriculture or branch of agricultural meteorology
Weather science	Includes agriculture and meteorology
Physical science	Biophysical science
Weather forecasting and other weather related service	Agroadvisory service to farmers

Figure 3.2. Difference between agro-meteorology and meteorology.

Source: Image by Wikimedia Commons.

3.5.1. Definition of Agro-meteorology

Agro-meteorology is a term which is abbreviated from agricultural meteorology and also, is referred to as agro-climatology.

- Agro-meteorology can be defined as "Science which investigates the meteorologic, climatologic, and hydrologic conditions, the conditions which are significant for agriculture owing to their interaction with the objects and the processes of agricultural production."

- Agro-meteorology can be defined as the study of those aspects of meteorology which have direct relevance to agriculture. Agro-meteorology is said to be the branch of science which investigates the relationship possessed by plants and climate.

Use of Agro-Meteorology in Crops and Animals

In crops, the agro-meteorology can be used:

- In regulating the plant cycle, growth, development, and yield;
- In giving the warning against the forest fires and cyclonic storm/ dust storm;
- In frost prediction and protection of crop plants;
- In controlling the insects and plant diseases;

The incidence of leaf miner in soybean is influenced by weather condition and not by the variety. As varying weather, infestation of leaf miner per cent varies and it may be controlled by controlling microclimatic condition with adopting the techniques for weather modification.

Sunanda Moka, M. G. Jadhav, P. R. Jaybhaye and B. V. Asewar. Effect of Weather Parameters on Incidence of Leaf Miner in Soybean. J. Agric. Res. Technol., 40 (2): 314-317 (2015)

- For minimizing the losses caused by heavy rain, floods, and hail storm;

The yield production model based on spectral reflectance and spectral ratios at the pod formation study of ground nut is helpful in assessing the crop yield one month in advance. Such information would be useful as discriminating features for remote sensing applications.

Jaybhaye P. R., M. C., Varshneya, and T. R. V., Naidu. 2002. Yield prediction model of summer groundnut based on spectral characteristics. Mausam, 53 (1); 53 - 56.

- In controlling the pollution of air and water;
- In soil and water conservation;
- Helpful in planning irrigation; and
- It is helpful in the modification of microclimate of crop fields which can be made to improve the health of the crop plants. This can lead to increase in the yield of the crop under modified microclimatic conditions.

The agro-meteorology is useful for animals in many ways which include:

- In designing animal houses: The houses for the animals can be designed in a scientific manner to save the animals from various climatic stresses like heat and cold waves which affect the production.
- It can be used for regulation of the animal cycle that is the growth, by-product, and yield (milk, eggs, and meat).
- It can be used in the prediction as well as prevention of animal diseases.
- It affects the growth of the pastures, and the health of the animals which depends on the selection of the pastures. The pastures are not good in cur climate but these are considered to be very good in the cool and temperate climates.

3.5.2. Agro-Meteorology as an Interdisciplinary Science

Agro-meteorology is an essential part of the agricultural system. This is an inter-disciplinary science. The study of weather and climate condition that results in the enhancement of the crops and the yield. This is an integrated system that constitute atmosphere, land, and ocean. It is also linked with the forestry, horticulture, and animal husbandry.

Along with the sound knowledge of meteorology, the agro-meteorologist requires a sound knowledge of agronomy too. The study of agronomy includes the study of physiology and plant and animal pathology, in addition to common agricultural practices.

The study of agro-meteorology involves:

- Atmospheric science;
- Soil science;
    - Plant science; and
    - Animal science.

Atmospheric and soil science involves physical and chemical environment.

Plant science and Animal science involves subjects like Pathology, Entomology, Parasitology, Physiology etc.

It is an interdisciplinary science. There are four stages in Agro-meteorology and the approach to deal with the subject:

  - Formulating accurate description of physical environment and biological responses;
- Interpreting the biological response in terms of physical environment;
- Weather and crop yield forecasts; and
- Control physical environment of crop fields and animal houses.

There are three aspects of meteorology; they are observation, understanding, and prediction of weather. The routine meteorological observations can be measured through various means. Some of the simple instruments like the thermometer for measuring temperature or the anemometer for recording wind speed (Figure 3.3).

Figure 3.3. Weather forecast.

Source: Image by Public Domain Pictures.

In recent years, the techniques have been emerging and is being complex in nature. Weather can be measured from satellites directly. The countries exchange information on the weather condition in minutes as the telecommunication has been an emerging technique.

Weather forecasts are made with the help of modern computers and super computers. These are plotted on weather charts which are then analyzed by the meteorologists at the forecasting center. The information on the weather forecast is of vital importance to many activities. The activities are agriculture, aviation, shipping, fisheries, tourism, defense, industrial projects, water management, and disaster mitigation.

Although the advances in the satellite and computer technology have improved a lot in providing forecast information to the users, still there is a lot left in terms of knowledge.

3.5.3. New Areas of Research

The rapid increase in the global population has led to the need for enhancing food production in the near future. This is achievable through the greenhouse experiments. Under controlled conditions, new models can be developed through various experiments. The areas where the agrometeorological research needs to be strengthened so that food security may be minimized are:

- The droughts and their ecological importance in agricultural production is needed to be studied.
- There is a need to compute the water balance index. To assess the crop production potentials of a given area.
- Under different microclimatic conditions, there must be study of the soils conducted for enhancing crop production.
- Experiments should be conducted in different agro-climatic regions for the selection of high yielding varieties.
- There is a need to conduct research on livestock under various climatic conditions.
- To develop agro-climatic models and dynamic simulation models for different crops, there is a need research. This should be done to predict the agricultural production before the harvesting of the crop.
- Research is also needed on the climate change. This is in context of global warming which is one of the burning topics. Many scientists of the world are concerned about the concentration of the green-house gases. Hence, it is important to give attention to the global warming.

There is an estimation that the temperature has increased by over past 100 years by 0.4°C. This climate change can result in affecting adversely the rainfall pattern over the globe. The change in climate is likely to affect the yield of the crop adversely in the years to come. So, in future, the evolution of new techniques is necessary for growing successfully under the climatic conditions.

3.5.4. Scope of Agricultural Meteorology

The agricultural meteorology is concerned with meteorological, hydrological, and pedological factors. These factors affect the agricultural productivity and it is concerned with agriculture's interaction with the environment.

The objective of this is to make such effects that help the farmers to apply supportive knowledge and information in agro-meteorological practices and services.

This knowledge extends from the soil layer of deepest plant and tree roots (pedosphere), through the air layer near the ground. This is the place where plants and animals live and crops grow. Then it extends to the higher levels of the atmosphere in which processes such as the transport and dispersal of seeds and pollen take place.

The field of interests includes the agricultural (including horticultural) production, forestry, animal husbandry, fisheries, and other forms of outdoor and indoor production, agricultural planning, processing, transport, and storage. This also moves to the agrometeorological components of food security, poverty reduction, and sustainable development aspects of the livelihood of farmers as well as of the use of their products.

The operational agricultural meteorology concerns itself with small-scale climate modifications in addition to the large-scale climate and its variations. For example, wind breaks irrigation, mulching, shading, and frost and hail protection.

There are certain other subjects for studying and that are agroclimatic characterization, pests, and diseases and their safe control, covered agriculture, quality of agricultural products, animal comfort aspects. The plant production for other than food purposes, including biomass as a renewable energy resource, and ecological considerations are some other to include.

There has been various monitoring done including early warning and estimation of changes of the risks of extreme events such as drought, desertification, and flooding. Much attention has been paid to the impacts of climate change and climate variability.

Specialized agriculture may turn unfriendly for the biodiversity. While on the other hand, intensive agriculture affects the environment through the generation of air pollutants, greenhouse gases (CO_2, methane, and nitrous oxide), ammonia, and tropospheric ozone.

The other modes that can cause soil erosion are through wind and water. Therefore, there is a major role of agricultural meteorology for understanding of emissions and pollutions from various unsustainable production systems.

Another key concept to devote attention to is the water management. This is to ensure that adequate supplies are there along with maintaining the quality of surface and groundwater.

The application of the subject to the aquaculture and fisheries range from the site climatology, hydrodynamics of rivers/reservoirs, estimation of contamination from agricultural run-off and of other ecosystem stresses to the usage of the meteorological factors. This is for the prediction of the occurrence of the toxic algal blooms.

There are research data, training, education research and the policy environments. These are the support systems to the agro-meteorological practices. Mathematical models are used increasingly in operational agricultural meteorology. This is mostly used in the industrialized countries with the Geographic Information Systems (GISs) to provide inputs to Decision Support Systems (DSSs).

The meteorological observations are being used by these models. However, there are the outputs of the operational numerical weather prediction ad of climate prediction. These forecasts can be used by the professionals or the decision makers.

Access to the additional biophysical parameters is providing by remote sensing, for example, vegetation indices and surface temperatures. Such data has been incorporated into the models. The agrometeorological information and services have led the training of the farmers and environment managers. This has turned the use of agro-meteorological practices and services as a matter of great importance.

3.5.5. Importance of Agricultural Meteorology

Except for the weather, there is no other branch of human activities that is dependent on agriculture. A large part of the agricultural production is dependent on the weather and climate. Despite various advances in the agricultural technology over the last half a century.

Climate change has led the agro-meteorological services to become essential due to the challenges in the agricultural production. Since there is increase in the climate variability and associated extreme events especially in the developing countries.

3.5.6. General Importance

There is knowledge in time and space of available environmental resources and the soil surface both below and the upper layer of the soil. The favorable and unfavorable guidance is provided for strategic agro-meteorological decisions in long-range planning of agricultural systems.

There are various examples of this guidance and the advancements in the agricultural activities. Such as designs of irrigation and drainage schemes, choices related to land-use and farming patterns, and within these choice selections of crops and animals, varieties, and breeds, and farm machinery (Figure 3.4).

Figure 3.4. Agricultural meteorology (1920).

Source: Image by Flickr.

In modern agriculture, the environmental issues are always dependent on each other, while the ecology and economy are on equal terms. More serious issues are emerging such as shortage of resources, destruction of ecological systems and other environmental issues. The uncontrolled use of chemical fertilizers and plant protection products disturbs the farmer's budget and is a burden on the environment.

In the short-term planning of the agricultural operations at different growth stages, there is a need to carry out detailed monitoring and real time dissemination of all the information related to the meteorology to the user of such information. This can be done by remote sensing, using satellites, radar, and other operational services. This will help in making tactical decisions for the planning of agricultural operations.

There should be a well-organized automatic production and coordinated dissemination of this information. Under this, tactical decisions imply decision in low external input sustainable (LEISA) agriculture, regarding the timing of cultural practices, such as plowing, sowing/planting, mulching, weeding, thinning, pruning, and harvesting.

It also includes high cost decisions, such as the application of water, extensive chemicals and the operation of costly crop-protection measures.

The farmers and other interested parties of the meteorological information should be able to make timely and efficient use of the information and other services of agro-meteorology. For this, the understanding of the effects of weather and climate on soils, plants, animals, trees, and production in farming systems, is necessary.

To achieve this, the right mix of the traditional adaptation strategies, contemporary knowledge in science and technology and appropriate policy environments is needed. Because without the policy support systems for agro-meteorological services, the production will always remain low or below the optimal range.

3.5.7. Applications

The accurate application of the knowledge of the weather and climate forecast depends on the availability of accurate and weather and climate forecasts or expected weather and climate patterns, depending on the time scale.

The accurate details of the weather forecasts range from the short-range weather forecasts (less than two days), medium range forecasts (MRFs) (less than 10 days) at certain critical times to seasonal predictions of climate patterns.

To keep the development plans safe and, in a position, that it can be used further, the indications of possible climatic variability and increasingly frequent and serious extreme events in the context of global climate change are of utmost importance as agro-meteorological services in addition to the application of other agro-meteorological information.

However, there is a shortage of reliable long-term weather forecasts relevant to the agricultural community on a routine basis all over the world. The agro-meteorological forecasts provide some of the significant services such as on the quantity and quality of crop yields, and the occurrence of animal and crop epidemics on the phenological events.

The established relationships of these forecasts are used between weather effects at an early stage of development and the final event. Forecasting is important for the assessment of crop conditions so that potential production anomalies may be recognized and quantitatively evaluated as early as possible.

The year to year fluctuations in weather patterns, as well as potential climatic variabilities and changes, on crop yields must be taken into account for long term planning. Surplus and deficit are always part of the crop yield. There are lots of factors that impact the global climate.

In the atmosphere, there are two most components that are CO_2 and water vapor. The forests that are beside the ocean absorb carbon dioxide and release water vapor. The burning of forests produces a large amount of carbon dioxide. Therefore, it is necessary to save the forests against fire and other human destructive activities such as destruction by insects, diseases, and pollutants.

The forest meteorology which works as a component to agro-meteorology provides useful information and services. The information for application to the forest authorities, foresters, and in case there is fire than to the fire brigades.

In developing countries, agro-meteorological services have greater responsibilities due to the increasing population pressure and the changing methods of agricultural practices.

In the future, there is an estimation of more and more demand that pertains to agro-meteorological information and services from the farming community. The demand relates to the technologies, farming system patterns, water management, weather-based pest, and disease control.

Therefore, future challenges are necessary for emphasizing the bottom up approach. As the information reaches even the small farmers such as the forecast information specific advisories and contingency planning in planning their day-to-day agricultural operations.

In developed countries, the agro-meteorological services mainly focus on the provision of environmental data and information to national policy and decision makers.

This helps in sustained food production, sustainable development, carbon sequestration in agro-ecosystems, and the land management practices that affect the exchange processes of greenhouse gases. This is because the developed countries might already have or developed technologies to adapt to climate change and climate variability, technology transfer might play a certain role.

However, local innovations remain the most important for application under very different conditions mostly in developing countries.

Advisories include:

- At various temporal and spatula scales, the information on average showing date as well as the expected sowing dates for the ongoing season and the operational crop protection of all the kinds.
- The information based on weather and agro-meteorological services in a specific location and user-friendly format. This information is used in more humid climates and the information includes on pest and disease attacks.

3.6. BENEFITS OF AGRO-METEOROLOGY

The agricultural productivity depends on the weather conditions. The growth and the harvest of the plants tell about the genetics and the environment surrounding the harvest. The careful planning and research by agro-meteorologist help farmers to meet the demand of the people for the food and other agricultural products.

The need for precise weather data has increased due to unpredictable weather patterns which is the result of climate change and other meteorological phenomena.

And this farmer pays close attention to rainfall. Too much rain might over export and remove the crops on the other hand two little rainfall main causes the crops to dehydrate and die.

The prediction of rainfall can save the crops as the farmers will come to know when to plant and water a crop. If there is too much rainfall then the farmers cover their crop with the plastic sheet and postpone the sowing of other produce.

A good example of this in the country Gambia, where the farmers to avoid this happening cover their crops with plastic sheets. There are many rigorous ways of accessing rainfall and link it to irrigation which involves calculating the water balance, estimating the infiltration of water within the crops and measuring evapotranspiration (ET) and soil moisture.

Soil moisture can be measured by using soil moisture and amp, and temperature sensor. Now this helps in knowing whether the soil has enough water retained in it to help the other crops grow. There are many other factors also that impact on agricultural operations other than rainfall.

The crops can be adversely affected by wind and humidity for example, forest fires. The burning and spread of fire can be prevented by the agriculturists through proper monitoring and observing these weather issues. This will allow animals to graze on unharmed crops that will provide more money to the economy.

In less endangering occurrences, the wind can also be measured. The endangering occurrences include strong gusts, in such circumstances the fixtures are needed to attach to the crops to save them and allowed them to stay upright.

There is a pre-basic practice that can be implemented in the day to day life of farmers which is planting crops near shady areas such as large trees. This will ensure that the plants are not overly exposed to sunlight and hence they will grow to their full capacity. This method of farming would result in an increase in the yield and produce a large quality harvest.

3.7. METEOROLOGICAL PARAMETERS IN AGRICULTURE

There has always been a challenge for the farmers in relation to the forecasting of the size and quality of the harvest. This is particularly for the cereal crops in relation to what has been sown. Also, there are many factors that have a direct and indirect impact on the final yield. However, the growth of any crop is not only dependent on the weather conditions alone yet they have a great impact on the final yield.

According to Dildar Hussain Kazmi and Ghulam Rasul 2009, in some countries the meteorological parameters are not vital for growth and development of wheat crop in rainfed areas throughout their growth period.

There has been a study on the wheat yield in which the effect of climatic parameters such as maximum temperature, minimum temperature, relative humidity (RH), Sunshine hours, and rainfall are studied (Chaurasia et al., 1991; Wardlaw et al., 1980; Rawson, 1986).

However, at some point in time these meteorological parameters act as the main actor for the physical growth of the wheat crop and the final yield. It is also experienced that at one stage or the other, the parameters of the meteorology had a definite impact on the physical growth of the plant.

Hussain and Mudassar (2004), has shown that in the mountainous region the impact of increasing temperature is always positive.

Pratley J (2003), shoes that the main factor behind the dry land yield fluctuations is the highest rainfall variance. Hoffman and Jobes (1978), has studied the influence of atmospheric RH on the growth of the plant and its interaction with salinity to influence the salt tolerance and water relation of barley, wheat, and sweet corn.

The influence of atmospheric RH and salinity phone the root crops like onion and radish has been studied buy Hoffman and Rollins (1971), and it has been found that the increase in humidity has not affected onion much while there is 15% growth yield in radish.

A study has been conducted by Tonkaz et al. (2010), to determine the effects of the change to the daily maximum-minimum and elevated carbon dioxide levels on winter yield of the wheat and yield components.

Wheeler et al. (1996), according to the study the yield of wheat has seen a decrease of 5 to 8% for every 1°C the rise in mean seasonal temperature. Mitchell et al. (1993), has an estimated decline of 10% at the end of the wheat for every 1-°C rise in temperature.

The meteorological parameters the most important to influence the growth development and yield of the crops are, Air temperature, RH, evapotranspiration (ET), wind speed, solar radiation soil temperature and rainfall.

At some particular stage of growth of the crops, these parameters are found to be more significant. For effective planning and execution of various agricultural engineering practices the examinations of these parameters can be done.

3.8. CONCLUSION

This is the chapter, which provides the basic detailing regarding the meteorology in context with agricultural meteorology. Meteorology is the branch of study, which deals with the atmosphere, atmospheric phenomena, and also, the atmospheric effects on the weather.

There are four scales of meteorology which have been very well described in the chapter which is named as mesoscale meteorology, microscale meteorology, synoptic scale meteorology and global scale meteorology. The history of meteorology has been mentioned in detail in the chapter.

The conclusion that can be drawn from this chapter is that agricultural meteorology is the branch of study which studies the atmospheric phenomena to agriculture and this branch is considered to be much useful in the field of agriculture (which includes crops and animals).

REFERENCES

1. Baier, W., Dommermuth, H., Samui, R., Shili, W., & Stigter, K., (n.d.). Importance of Agricultural Meteorology. [eBook] p. 2. Available at: http://www.agrometeorology.org/files-folder/repository/gamp2.pdf (accessed on 2 April 2020).
2. Boudreau, D., McDaniel, M., Sprout, E., & Turgeon, A., (n.d.). Meteorology. [online] National Geographic Society. Available at: https://www.nationalgeographic.org/encyclopedia/meteorology/ (accessed on 2 April 2020).
3. Huffman, G., (n.d.). Meteorology | Scholastic. [online] Scholastic.com. Available at: https://www.scholastic.com/teachers/articles/teaching-content/meteorology/ (accessed on 2 April 2020).
4. Munroinstruments.com. (2016). The Benefits of Agrometeorology. [online] Available at: https://www.munroinstruments.com/the-benefits-of-agrometeorology/ (accessed on 2 April 2020).
5. n.d. Agrometeorology. [eBook] p. 2. Available at: https://shodhganga.inflibnet.ac.in/bitstream/10603/155400/5/05_chapter%202.pdf (accessed on 2 April 2020).
6. n.d. Applications of Meteorology To Agriculture. [eBook] p. 2. Available at: http://www.wamis.org/agm/gamp/GAMP_Chap09.pdf (accessed on 2 April 2020).
7. Vijayan, P., Abdullah, A. T. D., & El Din, N. D., (2013). Significance of Meteorological Parameters in the Implementation of Agriculture Engineering Practices in and Around Tabuk Region, KSA. [eBook] p. 2. Available at: http://www.ijastnet.com/journals/Vol_3_No_5_May_2013/6.pdf (accessed on 2 April 2020).
8. Your Article Library, (n.d.). Agrometeorology: Definition and Uses. [online] Available at: http://www.yourarticlelibrary.com/agrometeorology/agrometeorology-definition-and-uses/88752 (accessed on 2 April 2020).

CHAPTER 4

Conceptualizing Radiation, Temperature, and Climate

CONTENTS

This chapter mentions the introduction to radiation, temperature, and climate. In the first half of the chapter, the readers will get insights about the concept of radiation and the application of radiation in context with agriculture. Furthermore, it includes details about the temperature and its effects on agriculture.

Along with this, the chapter also provides information about the concept of climate and its impact on agriculture. There is a concept of microclimate which has been well discussed in the chapter. In the second half, the chapter provides knowledge about the shift consumption patterns and the manure practices and livestock systems. In the last, the chapter provides insights about how the microclimates can be utilized in agriculture.

4.1. INTRODUCTION TO RADIATION

Radiation is defined as the process in which energy is emitted by electromagnetic waves or moving subatomic particle, particularly highly charged particles which cause ionization.

Emission of radiation from radioisotope decay can be effectively utilized in agriculture. Many problems regarding agriculture are addressed by scientists and researchers. These problems can be easily solved by means of radiation.

Apart from the harmful effects of radiation, they can be proved beneficial in terms of agriculture. These cannot be done by the means of traditional methods of agriculture. As modern problems require modern solutions, radiation is being effectively utilized in the agriculture field to enhance the crop yield, to preserve food, to determine groundwater.

Radiation is playing a vital role. Radioactive exposures are turning out to be an outstanding contribution by atomic energy as they lead to the improvement of agriculture. The adoption of radiation technology is continuously improvising agricultural productivity.

- **Animal Productivity and Health:** Various trans-boundary animal diseases are prevented by the means of radiation technologies. It also helps to increase livestock productivity. Milk yield of breeds like Bos indicus and Bos Taurus is increased by the farmers by the use of nuclear technologies from 500 to 1500 liters.

This helped them to generate an additional income of USD 110 million incomes per year. Brucellosis, a highly contagious disease which can transmit from animal to humans who drinks unpasteurized milk or undercooked meat from infected animals is curbed by the means of a program.

- **Improved Soil and Water Balance:** To ensure food security for the growing global population nuclear technologies are used in various countries for the maintenance of healthy soil and water system which are more important than anything. As an analogy under a scheme in Berlin 5000, rural farmers participated and outpoured an amazing result by increasing maize yield by 50% and lowering the amount of fertilizer by 70% along with the techniques that make nitrogen fixation easier.

Maasai farmers in Kenya adopted nuclear technologies to schedule small-scale irrigation, doubling vegetable yields while applying only 55% of the water. Earlier it would normally apply by conventional hand watering.

- **Pest Management:** Before allowing mass rearing over the pest-infested areas the male insects are sterilized by the means of nuclear-derived sterile insect technique (SIT). This technique is safer for the environment and human health than traditionally used pesticides as this technique inhibits and slowly discards already established pests or prevents the introduction of invasive species.

This technique of (SIT) is used by the governments of countries like Guatemala, Mexico, and the United States of America to desist the northward expansion of the Mediterranean fruit fly (medfly) into Mexico and the USA.

To protect valuable crops such as citrus fruits hundreds of millions of sterile male medflies are send by Guatemala to US states of California and Florida every week. This helps to control insect birth control as it unable to reproduce with sterile male medflies.

- **Food Safety:** To make an easier way of trade of safe food and avoid food fraud food safety and quality control systems needs to be vigorous at the national level, which costs the food industry up to USD 15 billion annually. By overcoming the problem of harmful residues and contaminants in food products and improving their traceability systems with stable isotope analysis up to 50 are benefited by nuclear techniques.

Testing of veterinary drug residues and contaminants in animal products such as scientific programmes started in Pakistan, Angola, and Mozambique are popularizing nuclear techniques. The new laboratory testing capabilities have helped 50 Pakistani food production and export institutes are benefited as they meet international food standards and helps to boost the country's reputation in the international food trade.

- **Emergency Response:** In the current scenario radioactivity exists in everything in our atmosphere. A sympathetic awareness of the movement of radioactivity through the environment becomes important to prevent or to make it less severe on agricultural products in case of a nuclear emergency. A large and strict obvious database on food contamination with radioisotopes was compiled together by FAO and IAEA in 2011 during a nuclear emergency in Japan.
- **Climate Change Adaptation:** Nuclear technologies are adapted by agriculture sectors for the adaptation of climate change by incrementing. The resource use ability and productivity in a sustainable way.

For helping farmers to breed, more productive and climate resistant farmers the nuclear derived cross breeding programme in Burkina Faso turned out to be an excellent example. The programme builds a strong foundation from the ground level by genetic evaluations in four national laboratories.

To produce a lick feed that provides broader and more productive livestock with nutrients required with more associated technology with nuclear technologies.

- **Seasonal Famine Prevention:** For ensuring food security, adapt to climate change and even to tackle seasonal famine vulnerable countries are being helped by crop-breeding programmes. The growing process is shortened by new mutant crop thereby helping farmers to grow additional crops during the planting season. Binadhan-7 a fast maturing mutant rice variety has been used recently by the farmers of Bangladesh.

This mutant rice ripens within 30 days comparatively much faster than the normal rice and this also provides extra time for farmers to harvest other crops growing in the same season along with these rice. The overall rice production of rice in Bangladesh has increased from 26.8 million tons in 2003–2004 to 33.8 tons in 2012–2013.

4.2. TEMPERATURE

The suitability of a crop to a particular region is mainly influenced by the climate being subdue factor. Climate largely influences the yielding capacity of a crop as well as the diversity of crop is determined by the climate. The temperature is one of the most important factors influencing the cropping pattern of a particular region has various effects. There are many more factors which influence the cropping system which include:

- Effect of temperature on plant growth;
- Temperature and crop production;
- Important for growth and development;
- Optimum temperature is required for maximum dry matter accumulation;
- High night temperature-growth of shoot; and
- Cardinal temperature points.

There are maximum, optimum, and minimum temperature limits of all plants. These are commonly called cardinal temperature points. Optimum temperature range is very important (Table 4.1).

Table 4.1. Optimum Temperature Ranges

Wheat	3–4°C	Minimum
	25°	Optimum
	30–32°C	Maximum
Rice	10–12°C	Minimum
	30–32°C	Optimum
	36–38°C	Maximum

Source: http://agritech.tnau.ac.in/agriculture/agri_agrometeorology_temp.html.

The temperature variation can give:

- Low temperature injury;
- High temperature injury.

4.2.1. Low Temperature Injury

Crop growth aspects such as survival, cell division, photosynthesis, water transport, growth, and finally yield are severely affected by low the temperature. Low temperature near the canopy due to the earth's re-radiation. If the cell size is large, the probability of frost damage is high.

Frost Damage

In colder regions due to incursion of large masses of cold air over region produces advective frosts.

- **Radiation Frost:** When the heat is freely radiated from all exposed objects on a clear calm night, this type of frost is termed as radiation frost.

1. **Hoar Frost (or) White Frost:** The frost which is caused due to sublimation of ice crystals on trees, branches etc.
2. **Black Frost:** The reduction of air temperature causes the freezing of vegetation.

Plant management against frost damage:

- Frost free growing season;
- Adjusting the sowing time;
- Selection of resistant varieties; and
- Sprinkler irrigation.

4.2.2. High Temperature Injuries

The temperature at which the plants die is termed as thermal death point. The annual temperature crop dies above 50-°C temperature. At lower temperature the shade loving plants are killed and the limit varies with plant.

The mineral nutrition, shoot growth and pollen development are adversely affected by high temperatures which leads to low yields.

4.2.3. Temperature Affecting Mineral Nutrition

- The reduction in absorption and subsequent assimilation of nutrients is caused due to high temperature.
- In maize the absorption of calcium is reduced at a temperature of 28 degrees Celsius.
- Air temperature and soil both affected by nutrient uptake.
- Under high temperature the nitrate reductase activity decreases.

4.2.4. Temperature Affecting Shoot Growth

- Temperate crops like wheat can be affected by high temperatures, even for a short period.
- The growth of roots and shoots are affected by high air temperature (Figure 4.1).

Figure 4.1. Shoot growth.

Source: Image by Pixabay.

4.3. CLIMATE

Agricultural sector is one of the most vulnerable sectors due to the climate change more than billions of farmers and their families face a great challenge as they solely depend on agriculture for their livelihood.

Climate is a dominant factor affecting agriculture. It determines the crop yield, biodiversity, water as well as soil health. Agricultural production is adversely affected by the climate changes. These changes occurred due to factors such as changing frequencies and intensities of precipitation, heat waves, and other extreme events which further leads to a rise in the prices of some major agriculture crops.

So, as far as these problems are being concerned there is an urgent need to implement certain solutions. So that these farmers and their families could get relief. There are still about 800 million undernourished and 1 billion malnourished people in the world. Although a lot of attention has been paid towards agricultural development and food security over the past years.

4.3.1. Impacts of Climate

There are many impacts of hydrological cycles. Due to the hydrological cycle more intense and frequent droughts and floods occurs in many agriculture regions. The crops can be destroyed and damaged due to the events caused by climate changes.

- **Heat:** Within the period of the next 30–50 years, the average temperatures will be supposed to rise by at least 1.0°C. The increased number of heat waves and warm nights, a decreasing number of frost days, and longer growing seasons in temperate zones are some of the anticipated regionally-dependent changes which will occur.
- **Carbon Dioxide:** Its concentrations will increase to about 450 parts per million by volume (ppm) over the next 30–50 years. C3 species (wheat, rice, soybeans), which accounts for 95% of world species are expected to have a larger effect on themselves as compared to C4 species (corn and sorghum). The elevated carbon dioxide levels are mainly responded by the C3 weeds, which shows the potential for increasing weed pressure and reduced crop yields.
- **Crop Biodiversity:** The climate change will adversely affect the distribution of wild crop relatives, which is an increasingly important genetic resource for the breeding of crops.
- **Economic Consequences:** Climate change will also lead to a rise in prices for most of the important crops such as rice, wheat, maize, and soybeans. This will further lead to the rise in prices of feed and therefore also for meat prices.

Such a scenario caused by climate change will lead to a reduction in the growth of meat consumption slightly as well as cause a more substantial fall in cereal consumption, which would further lead to larger food insecurity.

4.3.2. How Can Climate Change Impacts Be Predicted by Us?

The development of various climate models has been done. They also consider social and economic factors (population and carbon emissions).

Climate scenarios that express the potential for different behaviors to climate patterns are developed by combining all the above-mentioned factors with an understanding of global and regional climate science, by experts.

The impacts of temperature, precipitation, and seasonality on food production can be determined by these climate scenarios. These climate scenarios also allow both sufficient food production and food access, as the factors to rely on for food security. They can be defined as follow:

'All people, at all times, have physical and economic access to sufficient, safe, and nutritious food to meet their dietary needs and food preferences for an active and healthy life' (FAO, 1996).

Food access is currently acting as the principal barrier to food security. More then 10% of the world's population is undernourished although; sufficient food is globally produced to feed the current population.

The climate change mitigation increases the energy prices this may in turn lead to a rise in prices of food. Due to more frequent droughts and increased use of crop water, the amount of water required for food production may be scarcer over time.

As certain areas are becoming climatically unsuitable for production, the competition of land may increase. Sudden reductions in agricultural productivity which occur due to extreme climate events, which are associated with climate change, further lead to a rapid increase in prices.

For example, in the year 2010 areas including Russia, Ukraine, and Kazakhstan, suffered a rapid price increase of staple foods due to the heat waves in the summer season which in turn lead to yield losses in key production.

A sober demonstration of how climate change can lead to food insecurity is provided by the rising prices which are forcing a number of local people into poverty. The global mean surface air temperature will likely be increased by 0.4 to 2.6°C in the second half of this century (depending on future greenhouse gas emissions).

It is the substantial climate change that has already occurred since the 1950s. The Intergovernmental Panel for Climate Change has released this consensus. Agriculture and the wider food production system are already acting as a major source of greenhouse gas emissions.

Future intensification of agriculture is done for the compensation of production which is reduced (partly caused by climate change). It also increases the demand for animal products. All of this can further lead to an increase in these emissions. Between 2005 and 2050 it is estimated that the demand for livestock products will grow by +70%.

While in some regions, some crop yield can be increased if there is a gradual increase in temperature and carbon dioxide. They act as favorable conditions for increasing the crop yield. During the crop flowering time the potential yield, increases could adversely be affected by extreme events such as heat and droughts. Due to climate change crop production in many areas decreases during the 21st century.

A major challenge for agriculture which is likely to become more frequent are the heat waves (periods of extremely high temperature). A negative impact on food production in both plants and animals can be caused due to heat waves.

During the flowering time of plants if the extreme periods of high temperature can be proved dangerous. This may lead to no seeds at all if this single critical stage of plant flowering is disturbed. In the case of animals, the heat stress can lead to a decrease in productivity and fertility.

Negative effects of heat stress on animals also lead to weakening of the immune system which in turn makes them more prone to certain diseases. As the agriculture sector acts as the most vulnerable sector, it is likely to be affected by climate change adds a greater challenge for the farmers and their families.

4.4. MICROCLIMATE

It can be defined as the area which warmer, or colder, drier or wetter than the rest of the property due to its situation in the landscape. It is referred to as a local atmospheric zone which is slightly different from its surrounding areas.

It an area which is just a few square meters (for example a garden's corner) or it can also be large as many square kilometers (for example a valley).

The plants which are tender in your zone can be often grown just by understanding the theory which is behind the microclimates. More chilling required than the naturally provided by the climate for certain plants, these plants can be grown with the help of these theories.

Microclimatic study in canopy management of wine grapes showed that, the increased number of canes per vine decreased the canopy temperature and increased relative humidity. Influence of cane density on microclimate of canopy, growth, yield and quality of wine grapes N. S. HARWADIKAR, P. R. JAYBHAYE and T. B. TAMBE. Journal of Agrometeorology 15 (1): 79-81 (June 2013

If you think, or you find out yourself that taking up plants which are a few years old in your garden will thrive better in an area you can move those plant to that area. By this activity, experimenting and observation of your plants will let you will know more about the unique variation of your land in terms of soil moisture, light, and soil acidity.

With the help of this activity of experimenting, you will also help you to know about the plants which could easily adapt to the microclimate of your garden and you can assemble such plants.

4.4.1. Identifying Microclimates by Observing Moisture and Shade

There are five conditions which affect the microclimates and these are mentioned below:

- Temperatures;
- Patterns of light;
- Humidity distribution;
- Air circulation; and
- Soil pH and structure.

1. **Temperature:** People rarely are concerned about the temperature at which we can grow plants. It is 20 degrees at the thermometer when the thermometer says it is 20 degrees outside it can be taken as a key fact. Although the temperature is shown by the thermometer is 20°C but this does not mean this the temperature of your entire garden.

Furthermore, the variation of the thermometer can be noted or observed by taking it to a shady region and notice its changes. The one known as the sunny spots are the hot spots and they are invariable and possess the highest temperature. This is the area that is usually liked by the cats to lie around.

As it is also known by all of us that the hot air rises but keeping this in mind, we usually consider indoor heat and air balloons. Although this can be truly outdoor.

You can also experiment with this rise in warmer air by carrying the thermometer outside around your yard and can observe with an increase in temperature the air rises. The temperatures are slightly cooler in dips and valley., more sunlight or reflected heat is required by the more tender plants.

2. **Patterns of Light:** The amount of sunlight received by a particular area directly affects the microclimate of an area. The areas which tend to receive a large amount of sunlight are comparatively warmer to the areas receiving less sunlight. Below mentioned are some of the factors which directly affect the sunlight received by a particular area:

- Buildings;
- Soil mounds;
- Walls;
- Fences trees; and
- Other vegetation.

Shady areas are comparatively cooler than sunny areas of your garden and they tend to hold moisture for a longer period.

Different amounts of shade are created seasonally by some of these elements. In case, an area rapidly gets more sunlight than it did in summer might be due to trees that lose their leaves. In summer when the angle of the sun is higher about 75 degrees the shadows are shorter and the days are longer when compared to winters when the angle of the sun is lower (about 30 degrees).

The slopes which face towards south or southwest are warmer than the slopes which are facing towards the north or northwest when we take all these factors into consideration, such a pattern is observed. Such a condition usually occurs due to the direct receiving of the sunlight. As earlier mentioned, warm air rises and the cold air slides down the slope and it is settled at the lower points as they receive the sunlight directly.

A soil mound or stonewall (perhaps which is 2 meters high) facing towards the south can likely absorb the heat and light of the sun and then slowly will release the heat at night. This area can become a microclimate as much as 10°C higher than the other spots in your garden if this same area is protected from the winds.

With the help of shady place with more humidity and slight damper soil the plants which grow weaker in the hot summer can easily survive.

The one notable important fact is that the black light absorbs the heat while the light reflects the heat.

The shade trees can be put on the northern side of the garden for early spring or late fall is one of the ways to create a warm microclimate. The absorption of heat during the day and emitting it at night will in turn increase the amount of heat received from the sun.

3. **Humidity Distribution:** The amount of heat and cold in an area is also affected by water. A more moderate climate can be created by the presence of larger bodies of water nearby-such as the great lakes. A portion of the garden can even be affected by a small pond though only by small degrees.

The moisture is sent by lakes or ponds in the air. As the infrared radiations which are reflected from the earth's surface are trapped by the water vapors which acts as a bit like a miniature effect. We can get those reflected infrared rays, as long as there are vapors present in the air.

The heat is trapped by the air that takes on moisture from the pond. So, in this case some extra humidity will enjoy by the plants near the pond. During the day time the temperature there is warmer when recorded by the thermometer again. Humidity is also received by plants into the air.

This can also act as an advantage if we do the grouping at the edge of the plants which are drought tolerant and which requires higher humidity in the center of the bed in your garden. The humidity released by these drought tolerant plants around them can also benefit the plants which need more moisture.

4. **Air Circulation:** As mentioned, hot air rises, which means the reverse is also true-cold air sinks. Look around your garden early one spring morning or late autumn and see which areas have frost on the lawn. They are probably all the depressions in the land. It is easy to see that wherever the ground slopes downward will most likely be frosty-whereas the high points are a quite frost free. The point being: plants that need heat belong on higher ground.

The wind is another factor that affects the microclimate. Plants that are sheltered from winter winds survive better than those that are right out there getting the brunt of them.

Plant trees or shrubs in areas that get high winds if you need to create an area that is safe for those tender plants. The persimmons and almonds can be grown that are allegedly only marginally hardy in zone 7 and they are quite happy in our zone 6 gardens.

They can be grown in between a windbreak and a hedge of tall evergreen trees-and they are planted at what is the highest elevation on our property. They flourish for us because of the protection from wind as well as the heat and light reflected from the soil mound we have been creating over the years.

5. **Soil pH and Soil Structure:** Soil acidity is measured on the pH scale. A lower number equals the higher acidity, and because the scale is logarithmic, each number is ten times more acidic than the number above it. Neutral pH is 7, and plants vary in their tastes and tolerances.

Many traditional garden plants thrive the best with a soil pH between 6.5 to 7.0. You can learn the pH of your soil using testing kits that are easily found at garden centers. Check the pH of the soil in several different areas of the land, as it can vary considerably. Soil acidity is influenced by local sources of acidity (like many evergreens) or alkalinity (such as limestone).

Though there are many naturally occurring substances that you can use to change your soil's acidity, you can avoid a never-ending battle against your garden's characteristics by choosing plants that will enjoy the soil your garden offers.

Learning about the structure of your soil will also help indicate which plants will prosper. Soil structure can be determined by how it feels between your fingers. Sandy soil will crumb easily, clay will hold together like a ball when slightly moist.

The majority of plants will thrive in loamy black soil. Loam is a rich soil consisting of a mixture of sand and clay and decaying organic materials. It is quite high in nutrients to grow healthy plants (Figure 4.2).

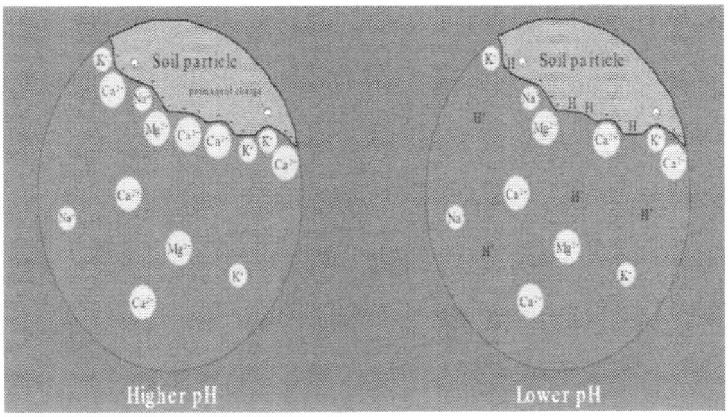

Figure 4.2. Soil PH.

Source: Image by Wikimedia commons.

For garden microclimates, think of moisture and shade. Using just these two factors can show you how to make a microclimate work in your garden. Each of the following is an example of a microclimate:

- **Dry Soil/Lots of Sun:** Plant drought tolerant plants. It is a great spot for a Mediterranean herb garden.
- **Dry Soil/Shade:** This is usually found under large trees. These areas may be cooler than the surrounding areas making them ideal for cool weather plants that wither in the sun.
- **Moist Soil/Lots of Sun:** Here's the spot for a water garden or bog garden. Plant anything that doesn't mind wet feet.
- **Moist Soil/Shade:** A woodland community. This is the perfect place for wild ginger, wild leeks, ginseng, goldenseal, black cohosh, blue cohosh, etc. For all those shade loving plants that love having their feet wet.

4.5. SHIFT CONSUMPTION PATTERNS

It will not be possible to reverse grow the agricultural emissions trends unless their root cause is addressed and that is the rising demand for agricultural products, particularly those that are carbon intensive.

Agricultural GHG emissions cannot be addressed simply as a problem of inefficient production on the supply-side. A spotlight must be cast on the pressures that inefficient, unsustainable consumption patterns pose to global climate and land use.

About 75% of this mitigation potential comes from changes in diet and the other 25% from reductions in food loss and waste. These major shifts in demand for agricultural products represent an emissions reduction of roughly 55% of direct agricultural emissions in 2030.

It is important to address rising meat consumption, particularly beef. Beef cattle represent 35% of direct agricultural emissions; dairy cattle and meat and dairy from other ruminants add another 28%.

Per unit of protein, or per calorie, beef, and other ruminants are extremely carbon intensive sources of food, even without considering cattle's role in driving deforestation. If global populations adopt U.S. consumption patterns, the associated emissions would be enormous.

Interventions that can help curtail major increases in beef consumption both in industrialized countries and in emerging economies will be critical over the next few decades. Given the established links between diet-related diseases and high levels of meat consumption, keeping global average per capita meat consumption at reasonable levels will have important health benefits as well.

In addition to dietary choice, several other interventions have promise including reducing the egregious levels of food waste and loss around the world and curtailing the use of food crops for biofuels. A range of actions including policy changes, behavioral change, and infrastructure investments can help address these issues.

Apart from the dietary choice, various other arbitrations have been included the reduction of energies levels of food waste and loss around the world. This also imposes a restriction on the use of food crops for biofuels.

These issues can be addressed by the range of actions which include policy changes, behavioral changes, and infrastructure investments. Major productivity gains can be achieved by focusing on key agricultural producers. To improve the efficiency of production demand side interventions need to be paired along with efforts.

The diffuse nature production is one of the greatest challenges in having the growth of agricultural GHG emissions. Although, GHG emissions can be reduced by a number of mechanisms but still there are only a few countries and sectors which can meaningful reductions (i.e., CO_2 reductions (Mt) per year by 2030 will be at least 40 to 50 million tonnes).

This can be done along with the practices which will be beneficial to the producers and yields. By combining several elements, the emissions reduction potential of the agriculture sector by supply based methods that account nearly 2 GT CO_2 per year by 2030. This also includes efficiencies gained in fertilizer production in China.

Nearly about 30% reductions from 2030 levels are represented by these emission reductions. India's dairy herd and reduction of enteric fermentation emissions from Brazil's cattle are the priority concerning areas that should be include.

Productivity gains are to in the best interest of farmers and governments as the mitigation opportunities are large. The quality of livestock diets is improved by opportunities in each case. This also helps the farmed animals to easily reach the market weight rapidly, which in turn results in more production of meat and milk.

The economics and productivity of the herds are improved due to these changes apart; they also result in lowering emissions per unit of product. A sustained production level is being supported to smaller animal populations by these changes. They also led to an increase in the efficiency of nutrient use on China's croplands. Globally, China is the country that has the greatest use of fertilizers. Without harming the yield certain measures can be applied to reduce GHG emissions from fertilizer application. Long term soil fertility and yield benefits would be provided by the reduced fertilizer applications in many cases. Traditionally China's fertilizer production has secured major industrial inefficiencies and that would yield very significant GHG reductions.

4.5.1. Reducing Rice Emissions in Southeast Asia

The opportunity to reduce rice emissions in Southeast Asia is spread across the region instead of a single country. The amount of rice grown in flooded fields has high potential of emissions and mitigation in rice farming.

Adding irrigation to better control water is one of the many interventions that have been used to reduce rice emissions and these are considered as complementary with productivity gains. This allows for double cropping.

4.6. IMPROVING STORED MANURE PRACTICES IN INDUSTRIALIZATION AND LIVESTOCK SYSTEM

The stored manure management does not benefit productivity in any way. There is also no serious food security risk and any other code and fats that the stored manure provides. While the mitigation interventions target these roads manual management. It has found that there is no other benefit provided by it.

In many countries, manure management is addressed through various progressive policies unlike many mitigation options. Every case should have an individual intervention. The intervention should be specific to country level conditions.

There are common interventions that are used for encouraging the changes in the agricultural practices these are expanding extension capacity, expanding the availability of subsidized loan, providing financial incentives, and working directly with groups of producers.

4.6.1. Pursuing Catalytic and Cross Cutting Interventions

Throughout the agricultural sector, achieving high productivity and low admission agriculture across the world will require that the mitigation practices become incorporated into day to day life of the producers.

In agriculture to adapt and create the best practices for an integrated climate and productivity agenda, the agricultural ministries agribusiness and other financial institutions and owners only to do this. Delivery job opportunities that are there in demand are already gaining attraction and is ought to be examined in more detail.

Schain agriculture investment is beneficial which steals money away from high emission agricultural activities to low emissions agricultural activities.

According to the recent commitment of the World Bank to climate smart agriculture, the opportunity may be particularly timely given. The climate oriented investments and commitments of major food and agribusiness would strengthen and ensures greater transparency and accountability in corporate supply chains.

Due to the jurisdictional limitations of intergovernmental, body both the United Nations Framework Convention on Climate Change (UNFCCC) and the World Trade Organization (WHO) the agricultural trade issues are stymied in both the organization proceedings.

In both the organizations that are WTO and UNFCCC, there may be incentives for agricultural mitigation measures that potentially removing barriers and break the grid lock. This can be done by targeted analysis.

In major agricultural economics, the reform of agricultural subsidies in particularly you and we would be enormously valuable. Although they are long term strategies advocacy then around this program would be worth the effort.

4.7. AGRICULTURE CARBON SEQUESTRATION: A RATIONAL APPROACH

The role of carbon sequestration agricultural soils and ground biomass has a game a lot of attention than any other agricultural mitigation in recent years. It has seen a lot of fluctuations as the discussion surrounding the role of carbon is increasing. As per BIS, report there is an estimate that global carbon sequestration potential would be between 700 and 1600 MT CO_2 per year by 2030.

The carbon sequestration practice is not well understood for mitigation yield and economic impacts viewpoint. There are also some complicating factors like the improvement in the nature of carbon stocks. As the challenges of agricultural carbon sequestration are, there so it should not be pursued in lieu of other mitigation opportunities.

However, it should be noted that the long-term management and preservation of soil carbon is very critical for the productivity in agricultural practices. This is because the soil fertility increases and there is a reduction in the soil erosion which indirectly increases the moisture in the soil.

The carbon sequestration is an important part of the climate solution in the agricultural systems. Farmers of all over the world need to maintain the organic soil matter regardless of the potential of it to measure it into the terms of money.

The carbon sequestration can be prioritized and supported by identifying those areas where soil carbon content is very low and the link to food security, poverty reduction and productivity gains are the strongest part.

4.7.1. Methods and Assessments

The integral part of CSA planning and implementation are assessment monitoring and evaluation. This factor is crucial for making decisions on the use of natural resources. The impact of climate change and variability on agriculture both past and future impacts and the vulnerability of communities are needed to be assessed.

To achieve goals related to food security climate change adaptation and mitigation or some other development objectives the effectiveness of CSA option should be assessed.

- MOSAICC

The Modeling System for Agricultural Impacts of Climate Change (MOSAICC) assesses the climate change impact on agriculture through simulations. There are many countries that carry out inter-disciplinary climate change impact assessment through this system.

- GLEAM

Global Livestock Environmental Assessment Model (GLEAM) helps in the simulation of the interaction of activities and the processes involved in the production of livestock such as meat and milk production and the environment. It is designed such that it can analyse multiple environmental impact categories including greenhouse gas emission, land use and land degradation, nutrient, and water use and interaction with biodiversity.

- SAFA

According to the guidelines of the Sustainability Assessment of Food and Agriculture (SAFA) system, the monitoring and evaluation of activities set baselines to define indicators measure progress and evaluate success in setback CSA interventions. SAFA is a holistic and inclusive framework for the assessment of sustainability performance in the food and agricultural sector and crop and livestock production along with forestry and fisheries.

- EPIC

The Economics and Policy Innovations for Climate (EPIC) smart agriculture program works with government, research centers, universities, and other institutional partners. It uses economic and policy analysis to support the transition to CSA. It is a work related program that aims at the identification and harmonization of climate smart agricultural policies analyzing its impact and effects cost and benefits along with the incentives and barriers to adapt the climate smart agricultural practices.

- Ex-Act

The Ex-Ante Carbon Balance Tool (Ex-Act) has been developed by FAO and is an appraisal system. During the development phase of the project, the ex-act estimates the impact of agriculture and forestry development projects programs and policies on the carbon balance. The carbon balance is the net balance from all the greenhouse gases which is expressed in

CO_2 and equivalent that will be admitted or sequestrated due to project implementation as compared to the usual business scenario.

- CRM

CRM is a short form Climate Risk Management. Addresses the vulnerabilities of the short-term climate variability and long-term climate change in the context of sustainable development. CRM is an integrated approach. The CRM provides usable weather and climate information products that help the farmer's livestock herders and fishes to assess the risk beforehand and improve opportunities at least the local level.

- Gender Mainstreaming

In a socially sustainable way gender, mainstreaming is crucial to achieving CSA. There is a need to understand that both men and women have access to do and benefit from CSR practices and policies. The role's responsibilities and capabilities are needed to be understood.

- Monitoring and Assessment of Greenhouse Gas Emissions and Mitigation Potential in Agriculture Project (MAGHG)

It falls under the MICCA program and supports the member countries in reporting and gathering data on GHG emission in the agriculture, forestry, and other land use sector for UNFCCC related reporting requirements. Fa analysis data for the improved that she's to respond to climate change and it assists many countries.

4.8. UTILIZATION OF MICROCLIMATES IN AGRICULTURE

The aspects of microclimates also affect plant ecology and are also applied to agriculture. Unfavorable microclimates can be avoided by planning well before planting the crops. This can be done by avoiding Frost pockets for sensitive crops and allowing for the effect of temperature on water balance.

However, the farmers can also try to make new microclimates which favor the plant in their growth process. The planted trees and bushes consist of shelterbelt that creates a drug that leads to slow down the drawing a cooling winds that blow across farmlands.

The effect of the shelter belt of trees on the wind speed can extend throughout the field which would be e more than 20 to 30 times the height of the tree.

Microclimate can be formed by greenhouses and other covers in agriculture. Earlier it was assumed that the main work of the greenhouse is letting the visible light in and preventing the infrared light from leaving the greenhouse. It is because glass is strong in absorbing infrared light (Figure 4.3).

Figure 4.3. Image of a microclimate.

Source: Image by World Atlas.

It is also known fact that the greenhouse mostly just heats up because they prevent heated air from rising away by convection as there is a roof on the top. However, this idea is no longer left in the science and now the greenhouse effect refers to the way certain gases in the atmosphere trap heat by letting the visible light in part blocking the way of red light from leaving the greenhouse.

In agro forestry, there is a creation of artificial shade and coolness of the forest under story. This is a tropical practice of growing crop plants under and between the rows of trees. When she did like this, the coca bushes do very well in the growth because they are descended from wild plants that naturally grow in the forest.

In California and Florida, if there is Frost then the citrus growers sometimes resort to radiant heaters in the open air of orchards. This is known to be a very efficient use of fossil fuel energy however it works very well that keeps the plant from getting affected by the Frost.

In Japan, the farmers resort to another trick to save their crops. The farmers of the rice usually increase their water level in the rice field to flood the plants with a layer of insulating water to protect them against a sudden cold spell. This is usually adopted when the most sensitive period is going on and which is when the rice is flowering and the Greens are beginning to form.

The solution to the improved crop productivity does not only depend on the rotational mono cropping system such as in the developed countries but it also depends on the polyculture cropping system which was traditionally used in the developing countries. This has led to the emergence of a new debate across the world about future agriculture at the time of food fuel and financial crisis.

Therefore, it is needed to Foster and agriculture which is inclusive, multifunctional, and which is built on principles of resilience that are crucial in the process of adapting to climate change (LEISA, 2008).

There is a need for such a farming system that increases food security decrease the environmental impact and responds to climate change. The enhancement of natural resource use provision of stability and high returns to the farmers are some of the changes that are needed. The agricultural benefits and environmental costs have seen a trade-off.

Hindi tropics and subtropics region there are the majority of agro forestry systems and practices that are presently still found. They are increasingly spreading into other regions also. The existing areas of agro forestry are certainly not of homogeneity as compared to the non-tropical forest.

Also, the distance between the trees is generally larger in agro forestry practices. However, the system is of a multi story agro forestry nature (Stigter, 1988, 1994).

4.9. CONCLUSION

There are certain factors that tend to influence the agriculture and the cropping patterns and they are radiation, temperature, and climate which have been mentioned in the chapter. There are a certain injury that can be caused due to the temperature variation that is high temperature or low temperature which in turn in effect the crops and agriculture in a negative way.

Climate is said to be a dominant factor that influences agriculture and so, can have positive as well as negative impacts on crops, cropping patterns and agriculture. Microclimate has also been discussed in the chapter. Microclimate acts as an element of air, sun, water, earth, and space.

The conclusion that can be drawn from this chapter is that there are factors that can affect agriculture in various ways which can be either in a positive way or in a negative way. Extreme cases of the factors can harm the crops but if used in a specified manner can make the crops flourish.

REFERENCES

1. Agritech.tnau.ac.in. (n.d.) Agrometeorology: Temperature and Plant Growth. [online] Available at: http://agritech.tnau.ac.in/agriculture/agri_agrometeorology_temp.html (accessed on 2 April 2020).
2. FutureLearn, (n.d.). Impact of Climate Change on Agriculture. [online] Available at: https://www.futurelearn.com/courses/climate-smart-agriculture/0/steps/26565 (accessed on 2 April 2020).
3. Goswami, S., (2017). Climate Change Impact on Agriculture Leads to 1.5% Loss in India's GDP. [online] Downtoearth.org.in. Available at: https://www.downtoearth.org.in/news/agriculture/climate-change-causes-about-1-5-per-cent-loss-in-india-s-gdp-57883 (accessed on 2 April 2020).
4. Kacera, S., (2011). Microclimates-[Veganic Agriculture Network]. [online] Goveganic.net. Available at: https://goveganic.net/article202.html (accessed on 2 April 2020).
5. Openweather.co.uk. (2017). The Influence of Temperature on Plant Productivity in Agriculture: Accumulated Temperature. [online] Available at: https://openweather.co.uk/blog/post/influence-temperature-plant-productivity-agriculture-accumulated-temperature (accessed on 2 April 2020).
6. Singh, B., Singh, J., & Kaur, A., (2013). Applications of Radioisotopes in Agriculture. [eBook] p. 1. Available at: http://large.stanford.edu/courses/2017/ph241/white-m2/docs/singh.pdf (accessed on 2 April 2020).

CHAPTER 5

Cropping Pattern and Cropping System

CONTENTS

This chapter is about the different types of cropping patterns and cropping systems that are being used to have a good yield to fulfill the need of each and every living organism.

This chapter includes different types of cropping systems. These cropping systems include, mono cropping, sequential cropping, intercropping, strip cropping, relay cropping and crop rotation. These systems are being used to grow one or more crops using these systems to have a proper and good yield of all the crops in a single field.

This chapter also explains the type of cropping patterns. This cropping pattern includes, mono cropping, mixed cropping, inter cropping and crops rotation. These patterns give the knowledge and proper guidance which helps in sowing two or more crop in a single field.

In this chapter, the main focus is one of the methods and different ways that are used to achieve proper and good yield.

5.1. INTRODUCTION

As there are many different combinations of crops in a farm and managing each crop combination in a different way. Fields have different sizes and different types of soil, which are helping in deciding which crop should be grown and help the farmers to choose the best crop for the field.

There are certain cropping patterns and some cropping systems that are used to have the best crop yield. Where cropping pattern plays an important role in planning the irrigation project. Cropping pattern helps in deciding the water requirement and also the capacity of the reservoir and canal system.

The crop variety is chosen according to the water availability, climate, soil texture, affinity, market, and income after harvesting, etc. in addition to the selection of optimal crop it is also important to improve the utilization potential of the irrigation projects that are already existing. So, this chapter focusing on the types of cropping patterns that are used nowadays for the irrigation command.

It is obvious that different crops are grown in different places, at different times. As there is the development of new varieties and hybrids, the crops are being introduced into new areas those offers a suitable environment and condition that are necessary for the growth of the crop.

Looking at the national map of the cropping pattern, on the western belt of great America is Nebraska that is considered as a corn belt. This Corn

Belt extends to the east through Iowa, Illinois, Indiana, and Ohio with the other major areas in the back of Texas to the southeast.

In the year 1920s and 30s, corn was considered as the major crop for Nebraska framers. They also grow oats, alfalfa, sorghum, and corn, which was used to feed the cattle and pigs.

Livestock management was considered as the main source of earning for the farmers. Farmers use to harvest the big crop and they use to sell some of the corn and grains that are necessary for the framers who need them as a feed.

Across the country, there are some factors that have changed the cropping pattern today. These factors are the development of hybrids, pesticides, fertilizers, new machines and irrigation patterns. In a country like America in the south during the 1920s and 30s, cotton was considered as a king along with the cattle and corn (Figure 5.1).

Figure 5.1. Image representing crops that are ready for harvesting.

Source: Image by PLOS Blog.

In the United States, cattle rising were the major activity of the framers across the country. But it was true that the highest concentration of the cattle was in Nebraska, Missouri, Texas, and Iowa. There is another breed that is used in most of the countries, known as hogs. Hogs are not much widespread as cattle but it is mostly found in Nebraska, Iowa, Indiana, and Illinois. These countries are considered as the center of the Hogs production.

This chapter is focusing the types of cropping systems and cropping patterns that are mostly used for the crops to have high yield.

5.2. CROPPING SYSTEM

Cropping system term refers to crops and crop sequence. It is considered as a management technique that is being used in a particular field over a period of years. The term cropping system is not a new word, but people started using this term recently when there was a discussion related to the sustainability of the agricultural production system.

There several other terms that are also been used during the discussion:

- **Allelopathy**: It is the chemical substance that is released by one plant species that usually inhibits the growth of another species. It has been proven that reduction in the yield can be caused when one crop follows another species of the same family.

For example, when a corn follows wheat. Auto toxicity is the damage that is caused to the crop when it starts following another crop. There are many cases where the actual reason for the reduction is not understood properly. But in general, it can be said that the reason for the breakdown of the crop residue can be due to release of the chemical that inhibits the growth of the crop that is to be grown next in the field.

So, to keep the old crop residue away from the new crop roots and seedlings should help to minimize such damages.

- **Double Cropping:** Which is also known as sequential cropping, is the practice of planting a second crop just after the harvest of the first crop. This is the process which helps in harvesting two crops from the same field in a year (Figure 5.2).

Figure 5.2. Image representing two crops peas and oats grown in a single field.

Source: Image by Progressive Diary.

This case is commonly known as a multiple cropping systems. This cropping system requires a season long enough and crops that get mature quickly and in a short time, that allows two harvests in one year.

- **Intercropping:** It is the presence of two or more crops in the same field at the same time. These crops are planted in such a way that their arrangement results in the crops competing with one another.
- **Mono-Cropping or Monoculture**: It refers to the presence of single crop presented in a field. Mono-cropping term is often used to refer to the growing process of crop that involves growth of a single crop in a field every year in the same field. This practice can be better described as a continuous cropping or continuous mono-cropping system.
- **Relay Intercropping:** It is a technique in which the crops are planted with different varieties at different times in the same field. With this both the crops are spent at least part of their season growing together in that field. For example, growing cover crops seed into a soybean crop before the crop gets mature.
- **Strip Cropping:** It is procedure where presence of two or more crops is planted in the same field. These crops are planted in a strip, which helps most of the plant competition that happens within each crop rather than between the crops.

Strip cropping is considered to have the element, intercropping, and monocropping. It consists of strips with particular width which determines the degree of each element.

- **Crop Rotation:** It is considered as a primary aspect of cropping system. It is said to receive the considerable amount of attention in the recent years. There are many people who are contending that current rotation is the unstable and even harmful for the environment and therefore it is considered as not sustainable.

There are many components of the sustainable agriculture that points towards the stability that is accompanied by the mixed farming practices that is done in the past.

In the mixed farming process, livestock is said to have the important role in utilizing the crops that are produced and in return, of the manure that are provided to the field. Such system is considered to still work well.

But this reduces the livestock numbers, fewer producer, and increase in the crop productivity that has meant to have that system. This type of system is likely to work well in relative to the small segment of the Illinois agriculture.

There is several types of cropping systems that are discussed in subsections.

5.2.1. Mono-Cropping

For example, planting maize crops year after year in the same field. Mono-cropping simply means that the field is used to grow one crop season after season. Mono-cropping is considered to have several disadvantages like it becomes difficult to maintain the cover on the soil because that encourages pests, diseases, and weeds. It also reduces soil fertility and it damages the soil structure.

So, it is suggested to avoid the mono-cropping technique. So, it better to used intercropping or strip cropping or rotating the crops every time after harvesting.

5.2.2. Sequential Cropping

For example, plantation of maize crop in the long rain and then planting beans during short rains. This system involves growing of two crops in the same field. These crops are planted one after the other in the same year.

In some of the places, the rainy season extends very long to grow two crops, which means that either two main crops are planted or one main crop along with the cover crop is planted. Plantation of two crops can be possible only if there are two rainy seasons or if there is enough moisture left in the soil to plant other crops as well. But if the crop is different, then that process is known as crop rotation.

5.2.3. Intercropping

For example, planting of crops in an alternative row with maize and beans, or planting the crop in between the cereal rows. Intercropping means growing of two or more crops in the same field at the same time.

This process can be done in different ways: spreading the seed of both the crops together or dibbling of the seeds with making any row arrangements.

This process is called as intercropping. This process is considered as easy, but it makes the weeding, harvesting, and fertilization process difficult.

As these plants are planted very close to each other then there is a possibility that plants may compete with each other. Plantation of main crop in a row and then spreading the seed of the intercrop or cover crop or when the plantation of both the crop and intercrop is done in a row. These processes are called row intercropping (Figure 5.3).

Figure 5.3. Image representing intercropping system used in field.

Source: Image by Dutch Openers.

The row arrangement makes the weeding and harvesting easier than the mixed intercropping. There is a possible problem that can arise is that the intercrop may compete with the main crop for the supply of light, water, and nutrients. Because of these reasons, this process may reduce the yield of both the crop.

5.2.4. Strip Cropping

For example, planting alternative strips of maize, soybean, and finger millets. Strip cropping is a process that involves planting broad strips with several crops in the same field. Each of these strips is 3 to 9 m wide. When this pattern is made on the slopes, then the strips can be laid out along the contour to prevent the erosion.

In this type of farming, farmers can rotate the crops by planting each of the strips with the different crop in the next year. Strip cropping has many advantages of intercropping, these advantages includes production of variety of crops, improvement of soil take place through legumes, and the rotation of crops helps in reducing the pest and weed problems.

The residue from strip can be further used as soil cover for the neighboring strips. At the same time, strip cropping has some of the disadvantages of intercropping that are being avoided by them, that is, managing the single crop with the strips is easy and there is no chance of having competition between the crops.

5.2.5. Relay Cropping

For example, plantation of maize crop firstly and then sowing beans in between the maize rows, which done after four weeks. So, relay cropping is the process of growing one crop and then planting another crop which usually works as a cover crop (Figure 5.4).

Figure 5.4. Image representing relay cropping that is done in a field.

Source: Image by Field Crop News.

This crop is planted in the same field before harvesting the first crop. This helps in avoiding the competition that takes place between the main crop and intercrop. It also uses the field for a longer time and as the cover crop usually starts to grow after the main crop is harvested.

5.2.6. Crop Rotation

Crop rotation is a process that uses one of the key principles that works for the conservation of agriculture. Crop rotation has many advantages, like, it helps in improving the soil structure as some of the crops have long and deep roots.

These long and deep roots can break the hardpans and tap moisture with the nutrients from the deep in the soil. Other crops are also there who may have many shallow roots. These types of plants tap the nutrients near the surface and also bind the soil.

These plants form many tiny holes because that will allow air and water to get into the soil. These plants help in increasing the soil fertility, as the legumes such as groundnuts and beans help in fixing the soil nitrogen. The nitrogen is released by the green parts and by the roots rot. This nitrogen can be further used by other crops such as maize.

This results in the higher and more stable yields that are without the need to apply expensive inorganic fertilizers. It also helps in controlling the weeds, pests, and diseases as planting the same crop season after season might have encouraged certain weeds, insects, and disease (Figure 5.5).

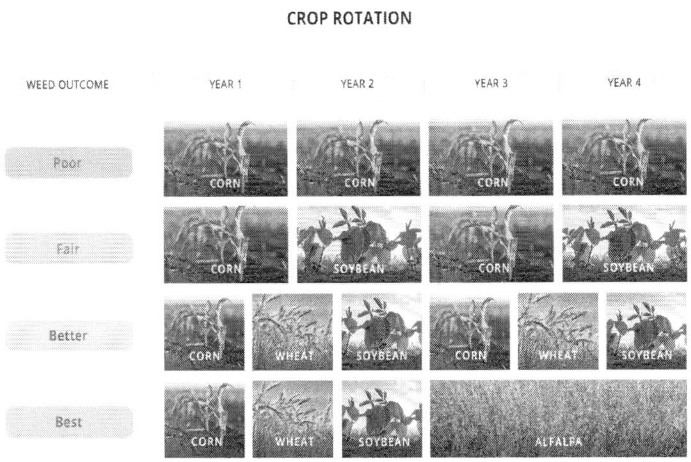

Figure 5.5. Image of example showing result of crop rotation.

Source: Image by Integrated weed management.

Planting different types of crops, helps in breaking the life cycle and it also prevents it from multiplying. When the mix of grains, beans, vegetables, and fodder are being grown, they produce different types of output that can consist of a varied diet and more types of product for the selling purpose.

The crop rotation reduces the risk as; single crop may fail because of soil conditions like drought. These plants have the chance that they may get affected by the pests and these conditions can make their market price fall, when time comes to sell those yields.

Therefore, producing or planting several different crops reduces the risk. There are some cases, where crop rotation takes the place of plowing the soil. This helps in aerating the soil, recycling the nutrients of the soil and even helps in controlling the weeds, pests, and disease. Intercropping, strip cropping and relay cropping techniques bring many of the same advantages as in crop rotation.

5.3. SEQUENTIAL CROPPING SYSTEM MANAGEMENT

Sequential cropping is different from the intercropping. In sequential cropping, crops are grown one after another and their management practices are event different.

5.3.1. Seedbed Preparation

As per the crop selection, suitable seedbed is being prepared. For example, preparing to puddle for the rice, development of ridges and furrows for the plantation of vegetables, maize, and cotton. Several flat seedbeds are used for other crops. However, there are two problems that are encountered in the seedbed in the sequential cropping system, these problems are:

- The available time for the seedbed preparation is considered to be less in high intensity cropping system. There is frequent rain that interferences the land preparation.
- To prevent the crop field also becomes difficult sometimes. For example, the preparation of field after harvesting the rice crop becomes difficult. As rice puddling method destroys the soil structure.

The time between the harvesting to sowing of next crop is considered to be more if the preceding crop is rice. To avoid this problem, minimum tillage or zero tillage should be adopted. It is considered as a common practice to sow the pulse crop just before or immediately after harvesting the rice crop.

In rice and wheat systems, the stubbles are being killed by spraying paraquat. Wheat is then sown in the plow furrows that are between the rice.

In case of irrigated agriculture, the time that is available for the tillage that takes place between the two successive crops is minimum which leads to minimum tillage.

Maximum tillage is often applicable on the soils with:

- A course texture surface soil;
- High biological activity of soil fauna;
- Good drainage;
- An adequate quantity of soil residue mulch; and
- Favorable initial soil moisture and friable soil consistency over a wide range of soil moisture.

In all sequence cropping systems, it is not possible to practice zero tillage or minimum tillage. For example, if the sunflower is the preceding crop, then plowing is said to be essential to oxidize the allelo-chemical of sunflower.

It is considered that the stubbles of pearls millet and sorghum contain the higher amount of Carbon Nitrogen ratio that immobilizes the nitrogen. Therefore, it is necessary to remove them. Stubbles are also considered to interfere the field of operations (Figure 5.6).

Figure 5.6. Image representing seedbed preparation by using machines.

Source: Image by Farm Equipment.

Whereas, in rice-rice-green grams system, summer plowing is done first. Later, in the rainy season when there is enough water availability then the puddling is done and crops are first grown. Second, the crop of rice is grown after minimum tillage. Green gram is sown as a relay crop at the time of second rice crop.

In case of cotton-sorghum-finger millet, the cropping system is done on the garden lands. Further, the field preparation is done and the field is laid out into the check basins to transplant finger millets. In the next season, another crop that is grown or planted among the stubbles of the finger millets is cotton which is done without preparing field. In the areas like Nigeria, inter-cultivation operation is controlling weeds.

5.3.2. Varieties

Short duration of crops is selected specially to fit it in the multiple cropping system. For a successful sequence cropping system, it is important and essential to have photo insensitive varieties. This needs most of the high yielding varieties.

5.3.3. Sowing

Sowing is not considered as a problem as there is sufficient time for the seedbed preparation. If there is case, where the seedbed is not prepared well there the establishment of the crops becomes difficult. For example, after rice, the cotton plantation becomes difficult in the black soil. As there is presence of hard pans in the shallow layer, which makes roots penetration difficult.

If the field is allowed to dry for the land preparation, then the sowing is delayed by one month. Therefore, the seedlings are being raised either on twisted paddy straw or the leaf cups. This whole process takes place before harvesting the rice crop (Figure 5.7).

Figure 5.7. Image representing sowing seed importance.

Source: Image by AFA journal.

After the harvesting of cow barn, a hole is made up to 30 cm this hole is partially filled with the sand and soil mixture, then in that the cotton seedling is being planted. The establishment of pulse crop becomes difficult after the rice crop.

When there is spreading of seed in the standing rice crop, then this will result in the uneven germination. Therefore, it is necessary to have the high seed rate. Crops are even planted in the stubbles are subjected to the competition by regenerating stubbles. It can be overcome by simply spraying, digging or preparation of stubbles.

Delay in the sowing is said to be a common problem in the intensive cropping system. When there is reduced in the yield which happens due to transpiration of the overage seedlings. With this, higher level of nitrogen is being applied to induce the tillering.

In the case of rice and wheat system, wheat yield is being reduced considerably, this happens when the wheat sowing is delayed beyond the November month. In this kind of situation, the planting is done with 40 to 50 days old seedlings of wheat are then used.

Soil fertility management is process which becomes more complex in the intensive cropping. This is because of the residual effect of the nutrients that is applied to the previous crops.

5.3.4. Soil Fertility Management

The soil fertility management possibly affects the legumes in the system with the complementary and competitive interaction that is achieved by the component crops, and effects the crop residues that are left in the soil.

There is some modern and chemical agriculture which includes higher cropping intensity which involves improved varieties, large amount of input of fertilizers and water. This helps in increasing the yields and accelerates in removal of plants nutrients that have added newer dimensions to the fertility management.

Fertilizers that are used for the sequence cropping: Based on long term fertility experiment that is conducted in various parts, below are the following broad conclusion:

- System productivity is being increased with the application of P along with N. There is further increment in the use of N, P, and K. the application of N is recommended when the dose is encouraged to each of the crops present in the cropping system.
- Phosphorous management in the cropping system needs careful adjustments of P fertilizers. The dose of these is taken into the account type of fertilizers. The soil characteristics and their yield level are being extent of the removal of P and for the growing environment.

In the cropping system, the involvement of wheat, P fertilizer dose to kharif crop can be reduced only if the preceding wheat has received the P in an adequate amount.

- In cropping system, the removal of K in the proportion to N is considered to be very high. This is particularly in case where cereals and fodder crops are involved. to maintain the soil fertility, it is very important to apply K fertilizers at the recommended dose.

- Among secondary nutrients, applying sulfur is considered to be beneficial particularly to the sesame-mustard, groundnut-mustard, and soybean-groundnut cropping systems (Tondon, 1991).
- Zn deficiency is considered to be most common among the micronutrients. Where 50% of the soil of intensively cultivated areas are suffering from the Zn deficiency. The crops that are grown especially in the rainy season are said to applied Zn more that the winter crops. Rainy season crops like maize, sorghum, and rice. Whereas winter season crops like wheat and chickpea. The use of organics is considered to prevent the Zn deficiency in the intensive cropping system that takes place under normal soils condition.

5.3.5. Cropping Systems and Integrated Nutrient Management (INM)

The concept of integrated nutrient management (IMS) involves the use of various types of inorganic, organic, and biological sources of the nutrient. They are required to improve and maintain the soil fertility which leads to the sustained crop production.

The crops responses to the organic and biological sources of the nutrients are not spectacular to the fertilizers. But the uses of these resources like supplementary and complementary uses are known to enhance the use efficiency of the applied fertilizers. Beside improving the soil physicochemical properties and preventing the emergence of micro nutrients deficiencies.

There are some major components of INM; these components are fertilizer, green manures, organic manures, biofertilizers, and crop residues. In case of cereal based cropping system, it is found that about 25 to 50% of the fertilizer that is NPK dose of the rainy season crops. This dose can get reduce due to the use of organics like FYM, green manures and crop residues (Figure 5.8).

Figure 5.8. Image representing the importance of maintaining nutrition in the soil.

Source: Image by TNAU.

In the system based on sugarcane, integrated sulphitation press mud, biofertilizers and cane trash are used. These are used with the inorganic fertilizers and green leaf manure shows the economy of fertilizers N about 20 to 25% that is to be applied to the sugarcane. This is applied by improving the use of efficiency of N, P, and other nutrients.

5.4. EFFECT OF INTENSIVE CROPPING ON SOIL PROPERTIES

Physical properties-there is continuous ground cover due to intensive cropping which is minimizing the soil erosion. They also lead to run-off losses and crust formation. There is improvement in the soil structure as there is relatively higher amount of crop residues which is due to intensive cropping.

Chemical Properties-it is important to maintain the optimum organic carbon in the soil. Therefore, recycling of organic residues presents in the intensive cropping system for the sustained production.

There are several factors that are used for determining the fertilizers schedule. These factors are discussed below:

- Nutrients added by legume crops;
- Soil supplying power;

- Residual effect of fertilizers;
- Crop residues left on the soil;
- Total uptake by crops; and
- Efficiency of crops in utilizing the soil and applied nutrients.

5.4.1. Soil Supplying Power

It is believed that growing different crops in a different season results in alteration of the soil nutrient status, which is being estimated by the soil analysis at the beginning of the season. There is an increase in the legume in the rotation when there is increase in the soil supplying power.

5.4.2. Crop Residues Left on the Soil

It is found that after groundnut, the amount of nitrogen and potassium that is available in the soil is higher than the initial status of the soil. But when the pearl millet is planted, after that only potassium status in the soil gets improved and there is no change in the P occurs.

5.4.3. Nutrient Uptake by Crops

The indication of the fertilizers requirement of the system is given by the total amount of nutrient that is taken up by the crops in one sequence. The balance is further obtained by subtracting the fertilizers that are applied to the crop to take the nutrients that are required by the crops.

5.4.4. Residual Effect of Fertilizers

It depends on the extent of residue that is left in the soil that what type of fertilizer should be used. There are some fertilizers like phosphatic fertilizers and FYM that have been considered as a residue in the soil.

These fertilizers are also useful for the subsequent crops. Whereas, the residue which is left by the potassium fertilizers are considered as marginal.

5.4.5. Legume Effect

It is considered that legumes help in adding nitrogen to the soil. This amount of nitrogen ranges from 15 to 20 kg/ha. The amount of nitrogen that is being added depends on the purpose.

When the green gram is planted in the field, it provides 24 and 30 kg of Nitrogen which play helpful role in the succeeding crop. When there is inclusion of leguminous green manure, then that adds 40 to 120 kg of Nitrogen/ha. The incorporation of green manure crops also helps in increasing the amount of phosphorous in the soil.

The groundnut crop residues also help in increasing the potassium availability in the subsequent crops. It is considered that the crop residues are adding considerable amount of nutrients to the soil. When the cotton crop is planted in the finger millet than the stubbles the benefits by 20 to 30 kg/ha. The stubble is caused due to decomposition of the stubbles.

It is considered that the deep-rooted crops are absorbing nutrients from the deep layers. For example, cotton, and red gram. The phosphorous gets added to the top layer, which is formed when the leaf falls and decomposes.

The crop residue is having high ratio of carbon and nitrogen. This is similar as in case of stubbles in sorghum, pearl millets which temporarily immobilize the nitrogen amount. It is also believed that the residue of legumes crop is containing low carbon and nitrogen ratio. They also get decompose quickly and release the nutrients.

5.4.6. Efficiency of Crops

For utilizing the nitrogen, jute is considered to be more efficient crop. Followed by the summer crops like maize, rice, groundnut, and potato in the order. Phosphorous efficient crops are jute, summer rice, kharif rice, potato, groundnut, and then maize.

Groundnut is considered as more efficient in potassium utilization that is followed by maize, jute, summer rice, kharif rice, and potato. It should be noted that the fertilizer recommendation should be on the basis of cropping system. For example, in case of wheat-based cropping system there is an extra dose of nitrogen that is 25% which is being recommended for the wheat when this is being grown after pearl millets and sorghum.

When the wheat crop is grown after the pulse crop, then the needed nitrogen per hectare is about 20 to 30 kg less. By using green manures, the phosphatic fertilizers are added, but not to apply phosphate to the succeeding wheat crop.

5.4.7. Water Management

It is believed that there is no carry over effect of irrigation when it comes in case of fertilizer. Rice is considered as efficient cropping system for the yield. But rice is considered to consume large amount of water especially in summers. In case of water scarcity in the summers, groundnut is grown instead of rice in the cropping system.

There are several methods of irrigation. The layout should be planned in such a way that it suits most of the crops. In case of rice-groundnut system; rice is being irrigated by the flood method, whereas, groundnut is being grown by the boarder strips. Whereas, in case of cotton-sorghum-finger millet system; were cotton and sorghum use furrow method and finger millet use check basin method (Figure 5.9).

Figure 5.9. Image representing water management done on field.

Source: Image by CCAFS-CGIAR.

There is presence of more remunerative and less water consuming crop rotation is being common at different places. The plantation of rice-mustard-green gram and rice-potato-green gram rotation is commonly found in most of the areas where there is presence of more water efficient system. Under high level of irrigation, the plantation of rice-lentil and rice-wheat cropping systems is found.

There are some crops like groundnut-rabi sorghum sequence was highly remunerative pre monsoon crops which use high water efficiency when compared to sugarcane when the irrigation water is not limiting.

5.4.8. Weed Management

Weed problem is often being observed in an individual crop. Weed shifting and carry over effects that are used as a weed control method on the crops that are needed to be grown after is usual. Weeds are considered as dynamic in nature. Generally, broad-leaved weeds occur in wheat at the later stage. Whereas, 2, 4 D is often applied as a post emergence herbicide to control them.

In case of rice-wheat system, canary grass is considered as danger for the wheat crop. It is considered that the seed of other species decomposes and they lose their viability, but the Phallaris minor seed does not lose the viability. But the Phallaris minor seed on the other hand does not lose their viability.

When sown in rice stubble, wheat is found to be heavily infested with the Phallaris minors. In case of zero till, cotton-sorghum-finger millets the weeds growth is being controlled by using the herbicides. Herbicides application can be found toxic when applied to the previous crop. Herbicide will make the succeeding crop toxic.

When the higher dose of Atrazine is applied to the sorghum crop, this may affect the germination of the succeeding pulse crop. Herbicides recommendation should be on the basis of succeeding crops. Plowing before planting helps in killing most of the weeds.

5.4.9. Pest and Diseases

Pest and disease invasion are more in sequence cropping. This invasion is due to continuous cropping. Carry over effects of insecticides is not being observed.

5.4.10. Harvesting

In the sequence cropping, instead of harvest maturity crops are being harvested at a physiological maturity stage. The field can be vacated one week before. If there is continuous cropping, then the harvesting may get coincide with the heavy rain and with the specific post-harvest operation.

These post-harvests may include artificial drying, treatment of crop with the common salt, etc. these are some of the practices that are being done to save the production.

5.5. CHALLENGES IN CROPPING SYSTEMS

There is some problem that can be easily seen with the cropping system. But there are some ways to overcome them. There is a possibility that certain pests, insects, and diseases my get spread easily from one crop to another via crop residue. In this case crop combination should be avoided.

There is a possibility that markets are not present always for the new crops, then planting is considered as a part of once rotation. It can be possible that the person may find for seeds and they are not interested to buy the yields or the crop value is too low that becomes useless of growing crops.

It is important to check the source of the seeds and price of the final product before deciding to crop the plant. There is only one possibility that these crops are being sold at bigger markets. While managing rotation properly, it requires more skill and knowledge than a single crop need. People may be very excited to try out new crops because they are not using for the purpose of growing or feeding. New crops should be practiced on a small scale and in a small area so that the correct information regarding the good growth of that crop should be known properly. It is also important to visit those farmers who are already growing these crops and ask their working advice to have proper information regarding that crop before growing it or sowing it.

Cropping pattern refers to the proportion of land that is under cultivation of different crops at different points of time. Cropping patter indicates the time and arrangement of the crops in a particular land area.

5.6. CROPPING PATTERN

If there is any change in the cropping pattern, then that can cause:

- Change in the proportion of land under different crops; and
- Change in space sequence and time of crops.

In many areas, cropping pattern is being determined by the rainfall, climate, soil type, technology, and temperature. To obtain maximum yield, different patterns of cropping are being practiced. There are some major cropping patterns that are explained below:

5.6.1. Mono-Cropping

Mono-cropping is process that is reducing the soil fertility and also destroying the structure of the soil. So, for the upgradation in the production there is requirement of chemical fertilizers. This practice allows the spread of pests and diseases.

5.6.2. Mixed Cropping

Mixed cropping is a process that involves two or more crops that are grown on the same land simultaneously. For example, growing of wheat with the gram on the same field at the same time is known as mixed cropping. This practice minimizes the risk of failure of one crop and also ensures that the crop failure may be due to the abnormal weather conditions.

The crops that are chosen for growing together should have a different maturation time and different water requirement. It is considered that one tall crop and one short crop should be grown together. Their requirement should be different, as one crop should require fewer nutrients than the second one (Figure 5.10).

Figure 5.10. Image representing mixed cropping patter on field.

Source: Image by Quora.

It should be also kept in mind that one crop should have deep roots capacity while other roots should be shallow. All these criteria are very important to have a successful mixed cropping pattern.

There are some advantages of mixed cropping that are explained below:

- There is an increase in the crop yield;
- The infestation in the pest is minimized;
- There is a reduction in the risk of crop failure;

- There is a proper utilization of soil; and
- As more than one crop is planted, therefore more than one variety of crop can be harvested at the same time.

5.6.3. Intercropping

Intercropping is the process where more than one crop is grown in the same field and even at the same time in a fixed row pattern. In this row pattern, first row is of the main crop then, next three rows are for the intercrops. In this way, it helps in increasing the productivity of per unit area. There are many advantages of intercropping that are being discussed below:

- In terms of yield per unit area, intercropping gives more yield than the sole cropping.
- It behaves like an insurance that works against the failure of the crop in the abnormal years.
- Intercropping helps in maintain the soil fertility. This is because, it helps in taking the nutrients that are made from both the layers of soil.
- Intercropping helps in the reduction of soil runoff and it also helps in controlling the weeds.
- Intercropping provides shade and support other crops in the field.
- It is proved that intercropping system utilizes the resource efficiently and it also helps in increasing the productivity.
- Intercropping when done with the cash crops are considered to be higher profitable.
- It also helps in avoiding the intercrops competition. This directly helps in growing higher number of crops plants per unit area.

Intercropping system can be of different types:

- **Row Intercropping:** It is a system in which components of the crops are arranged in an alternative row. This type of cropping system helps in optimizing the utilization of land space and it suppresses the growth of the weeds during early stage of main crop.
- **Strip Intercropping:** This system is a process in which two or more crops are grown in a wide strip. This pattern helps to manage the crops separately. Therefore, this makes the crops close enough to each other to interact easily.

1. **Relay Intercropping:** It is the type of intercropping in which second type of crop is being planted when the existed crop is ready for flowering but not ready for the harvest. For example, rice-cauliflower-onion-summer gourds.

5.6.4. Crop Rotation

Crop rotation is the process in which different crops are grown on the same land in the pre-planned succession. Depending upon the duration of the crop, the crops are being classified as one-year rotation, two-year rotation, and three-year rotation.

To increase the soil fertility, the legumes are being included in the crop rotation programme. The crops which require high fertility level that can be grown after the legumes., the crops that require high inputs are grown first then the crops which require low input.

The criteria that are used for selecting the crops for crop rotation:

- Sufficient amount of moisture should be available;
- There should be enough availability of fertilizers, man power and machine power;
- There should be easy availability of marketing and processing facilities;
- There should be enough availability of nutrients in the soil; and
- The duration of the crop should be short or should be long.

Advantages of Crop Rotation:

- This process helps in maintaining the soil fertility for a long period;
- This helps in preventing the growth of weeds and pests;
- This process doesn't require lots of chemical fertilizers; and
- This process helps in maintaining the physical and chemical nature of the soil which remains unaltered.

5.7. FACTORS AFFECTING CROPPING PATTERNS

There are several factors that are affecting the cropping pattern. As the cropping pattern helps in determining the level of agricultural production. This reflects the agricultural economy of any region.

The cropping pattern is affected by the changes that occur in the agrarian policy, the availability of the agricultural inputs and the improvement in the technology. Therefore, the cropping pattern is very beneficial for improving the fertility of the soil. It also helps in increasing the yield of the crops. It directly ensures that crop protection and availability of the nutrient in the crops.

Crop rotation is a process that involves planned sequence of the growth of different crops in the same field. Crop rotation is opposite of the continuous cropping, which is grown in the same field with the same crop year after year. Ideally, it meant that these crops are not of the same family.

5.7.1. Why Do We Rotate Crops?

Crop rotation is mostly practiced on a sloping soil. As sloping soil has the potential for saving soil. Crop rotation is used to improve or maintain the conditions of the soil. These conditions can be physical, chemical, and biological. Crop rotation is also used to reduce the average rate of soil erosion from a field.

To reduce the soil erosion and also improve the soil structure, the grasses and legumes are included in a rotation pattern that works very effectively. When there is a use of legumes in the field in the rotation then it helps in eliminating the need for the nitrogen fertilizers.

There are other crops that accumulate the phosphorous or potassium. In addition to above, crop rotation can also work as an important part of an Integrated Pest Management Program (IPM).

Other values of crop rotation are explained below:

- **Plant Nutrition:** As it is already known that each of the crop uses different types of minerals and in a different amount from the soil. It is said that after sometime the soil starts to get depleted of the minerals that are essential for the plant growth and health if the same crop is grown each year.
- In return, different crops will sometime return the missing minerals to the soil. This happens when the plant dies and starts to decompose or it get mixes into the soil.
- **Soil Structure:** Crop rotation helps in preserving and improving the structure of the soil. Crops are considered to have different roots structure and they grow even to various depth. By rotating

the soil means deep diggers in the soil that will slowly deepen the top soil.

- **Insect Control:** Insects controlling is very important as they over the winters in the soil. The insect gets to enter in the leaves and vines of the plants. They make themselves ready for the reawakening in the spring season to find their favorite meal. These insects face the plant, they do not feed those plants.
- **Disease Prevention:** Similar to the insects, the disease prevention in the plants is very important. Plant disease is responsible for over ring the winters in the plant leaves, vines, and roots that are under the soil. Crop rotation helps to guard against these diseases.
- **Water Quality:** It is important to maintain the water quality of surface water as well as ground water. Surface water quality is being improved by simply reducing the sediment loss and reducing the losses of dissolved and sediment that have attached nutrients and pesticides.

The loss of nitrogen from the ground water can be reduced by deep rooted sod crops. These crops can use the nutrients from the deep of the soil profile. Legumes crops help in fixing the atmospheric nitrogen. That in turn can reduce or eliminate the need for commercial nitrogen fertilizer for the following crops.

It is believed that crop rotation also helps in encouraging the healthy root systems. This will work effectively by retrieving the nutrients from the soil. Therefore, this helps in minimizing the leaching to ground water.

5.7.2. What Is Process of Rotating the Crop?

The crop should be rotated on at least three years or four year of cycle. The crops should be rotated every year. For example, if the corn crop is being planted this year, it should not be grown or planted in the same field next two years or three years.

Therefore, ideally different crops should be used every year in the same field, as this helps in decreasing the possibility of growth of the insects and diseases. This helps the crops to get safe from the effect of the insects and disease and in this way, the crop will likely affect the similar crops. For example, cabbage, and broccoli, both are from the same family and they both should not follow each other. Crops should be changed year by year in a planned sequence.

Planning Consideration includes the pattern which emerges when the best rotation plan is chosen. Legumes are considered to be beneficial in preceding crops. For example, crop of potatoes is believed to yield best when planted after corn.

Some of the preceding crops like oats, barley, and peas are said to increase the frequency of scab on the potatoes plant. Some crops like Corn and beans are generally considered that they do not greatly influence by the preceding crops. Whereas, crops of carrots, cabbage, and beets are generally considered to damage the subsequent crops.

When there is a use of legumes in the crop rotation, the nitrogen that is formed by the fixation should take into the account. While determining the nutrients that are needed by the future crops, the over application of nitrogen should be avoided and should be prevented.

Therefore, it is important to monitor the soil fertility on the regular basis and the fertility that is maintained should be within the acceptable range for all the crops used in the rotation.

While planning out a rotation process, the crops are needed to be divided into their families. This follows the principle which includes, not growing out the same crop or one crop of the same family. For example, beets, spinach, and chard are in the same family. It is also important to note the proper space required by the crop. For example, crop of radish is considered to require much less space than the corn crop.

In addition to rotating crops, there are many farmers who are rotating their livestock among different sections of the pastures. This helps in the diffusion of manure in the fields as well as preventing the over grazing of any section. Overgrazing of any of pasture can lead to the depletion of the vegetation and resulting in soil erosion.

There are some points that are very important for the crop rotation:

- Crops that are used for this purpose should be suitable to the soil;
- Small grains or meadow helps in providing better erosion control in the crop rotation;
- Any of the low residual crop or soybean crop can be replaced by the corn during rotation to gain better erosion control; and
- Plantation of grass after grass should be avoided.

5.8. CONCLUSION

As there are several methods that are being used as cropping systems and cropping patterns. These crops are being grown, to feed the people around who need them. Agriculture deals with crops and related fields. To have a proper yield of crops there are certain criteria that are being used to have good amount of yield.

Good yield is important, as this is the basic requirement that is increasing day by day. To fulfill the need of each and every living being, agriculture plays a vital role.

To have a good yield there are certain cropping patterns like monocropping, intercropping, crop rotation, and mixed cropping. These patterns are being used to achieve a good yield and proper yield.

Similarly, there is use of certain cropping systems which are being used to grow lots of crops using the techniques. These systems include, mono-cropping, crop rotation, sequential cropping, intercropping, strip cropping, relay cropping. These cropping systems explain the ways through which different type and varieties of crops are grown in a single field.

There are also some effects that are impacting the soil properties which involves intensive cropping system. This topic explains about the reason which is responsiblefor degrading the quality of soil which directly impacts the growth of the crop.

REFERENCES

1. Civilsdaily.com. (2017). Types of Cropping Systems: Mono Cropping; Crop Rotation; Sequential Cropping; Inter Cropping; Relay Cropping- Civilsdaily. [online] Available at: https://www.civilsdaily.com/types-of-cropping-systems-mono-cropping-crop-rotation-sequential-cropping-inter-cropping-relay-cropping/(accessed on 2 April 2020).
2. GKToday, (2015). Types of Cropping Patterns. [online] Available at: https://www.gktoday.in/gk/types-of-cropping-patterns/(accessed on 2 April 2020).
3. n.d. Crops and Cropping Systems. [eBook] Available at: http://www.act-africa.org/image/06CROP~1.PDF (accessed on 2 April 2020).
4. Nafziger, E., (n.d.). Cropping Systems. [eBook] Illinois. Available at: http://extension.cropsciences.illinois.edu/handbook/pdfs/chapter05.pdf (accessed on 2 April 2020).
5. Rana, S., & Rana, M., (2011). Cropping System. Research Gate, [online] Available at: https://www.researchgate.net/publication/309211205_Cropping_System (accessed on 2 April 2020).
6. Reinhardt, C., & Ganzel, B., (2003). Cropping Patterns | Wessels Living History Farm. [online] Livinghistoryfarm.org. Available at: https://livinghistoryfarm.org/farminginthe30s/crops-3/cropping-patterns/ (accessed on 2 April 2020).
7. Sustainablesettings.org. (n.d.). Crop Rotation | Sustainable Settings. [online] Available at: http://sustainablesettings.org/crop-rotation/ (accessed on 2 April 2020).
8. Toppr.com. (n.d.). Cropping Patterns | Definition, Examples, Diagrams. [online] Available at: https://www.toppr.com/content/concept/cropping-patterns-200426/(accessed on 2 April 2020).

CHAPTER 6

Planting Geometry and Its Effect on Growth and Yield

CONTENTS

Plant Geometry is the space and the shape of the space available for the individual plants. This chapter discusses all the plant geometry and the various aspects related to it. The first half of the chapter explains the meaning of plant geometry, different methods of sowing and transplanting. In the mid of the chapter, the plant population and the factors affecting plant population are explained. Later in the chapter, various equipment that helps in the plantation of the crops are discussed.

6.1. INTRODUCTION

6.1.1. Plant or Crop Geometry

Plant or crop geometry is the shape of the space available for individual plants. The crop geometry influences the yield of the crop through its influence on light interception, rooting pattern and moisture extraction pattern. Crop geometry can be changed by changing inter and intra-row spacing or the planting pattern.

Wider spaced crops have advantage under this geometry:

- Plants which requires no restriction in all directions are given square geometry.
- Mostly, there is perennial vegetation like trees and shrubs which falls under this arrangement.

Crop geometry is concerned with the arrangement of the plants in different rows and columns in an area to utilize the natural resources efficiently. In other words, area occupied by a single plant is the crop geometry.

For example, Rice – 20 cm × 15 cm. This is very essential to utilize the resources like light, water, nutrient, and space. Different geometries are available for crop production.

6.1.2. Square Plant Geometry

Under this method, the plants are sown at equal distances on either side. Mostly perennial crops, tree crops follow square method of cultivation. Ex. Coconut – 7.5 × 7.5 m; banana – 1.8 × 1.8 m. But, due to scientific invention, the square geometry concept is expanded to close spaced field crops like rice too.

The arrangement of plant in the form of the square is more efficient in the utilization of light, water, and nutrients available to the individual plants than in a rectangular arrangement.

Advantages:

There is uniformity in the light, movement of wind is not blocked and mechanization can be possible.

6.1.3. Rectangular Method of Sowing

Under this method, there are rows and columns, the row spacing is wider than the spacing between plants. The different types of rectangular method are:

The sowing of the crop with seed drill, wide inters row closer intra-row and closer intra-row spacing leads to rectangularity. The ease in the cultivation is facilitated by rectangular arrangement. Rectangular arrangement facilitates easy inter-cultivation. Rectangular planting mainly suits annual crops, crops with closer spacing etc., the wider section (row) is given for irrigation, intercultural operation etc.

- The arrangement that restrict the endless growth habit to switch over from vegetation to the productive phase.
- High density planting is possible in this arrangement.
- Intercropping is also facilitated by this.

6.1.4. Solid Row

There will be no proper spacing in the rows between the plants. This is followed only for annual crops which have tillering pattern. There is a definite row arrangement but no column arrangement, For example, Wheat.

6.1.5. Paired Row Arrangement

It is also a rectangular arrangement. Under this method, a crop requires 60 cm x 30 cm spacing and if paired row is to be adopted the spacing is altered to 90 cm instead of 60 cm to accommodate an intercrop. However, the base population is kept constant.

6.1.6. Skip Row

As the name suggests, a row is skipped and hence there is a reduction in population. The reduction in the plant population is compensated by planting an intercrop; practiced in rain fed or dry land agriculture.

A row of planting is skipped and hence there is a reduction in population. This reduction is compensated by planting an intercrop; practiced in rain fed or dry land agriculture.

6.1.7. Triangular Planting

Triangular cropping is the method to accommodate plant density under perennial/tree crops. It is done by joining the three points of the crops with the help of an imaginary line. It is recommended for wide spaced crops like coconut, mango, etc. The number of plants per unit area is more in this system.

6.1.8. Miscellaneous Planting

The plantation of rice and ragi is done either in rows or at random. A skip row planting is done by skipping every alternate row. The decreasing inter row spacing is adjusted when the density of one row is skipped.

The reduction in the inter row spacing between two rows and the increase in the spacing between two such pair is known as paired row planting. It is generally done to introduce an inter crop.

Under the random plant geometry, broadcasting method of sowing is the basis for the sowing and no equal space is maintained in this. Resources are either under-utilized or over exploited.

6.2. METHODS OF SOWING AND TRANSPLANTING

6.2.1. Broadcasting

It is another name for random sowing of seeds. The literal meaning of this method means "scattering the seeds." Broadcasting method is adopted for many crops. Generally, this method is adopted for small sized to medium sized crops. This method is mostly followed in India. This method is the easiest and the cheapest which requires minimum labors (Figure 6.1).

Figure 6.1. Seeds sown through the method of broadcasting.

Source: Image by Wikimedia Commons.

To have the optimum plant population in unit area, there are certain rules to be followed:

- The scattering of seeds should be uniform, and it can be achieved by a skilled person only.
- to trigger the germination, the plowed field should be in a perfect condition.

The sowing is completed strip by strip as the broadcasting is done in a narrow strip. It is advisable to broadcast on either direction for ensuring the uniform plant population. This is called crisscross sowing. In case a seed is too small, it gets mixed with the sand to make a bulky seed which is easy to handle. For example, Sesame seeds are mixed with sand at 1:15 or 1:10 ratio and sown.

There are cases when the person who is sowing the seeds has to beat the seeds against the basket for uniform scattering. Ex. Sorghum, pearl millet. After the broadcasting is done, the seeds are covered gently by using a country plow with a very shallow plowing or some wooden planks (boards/levelers).

In some cases, tree twigs or shrub branches are used. In case of large seed, the leveler collects the seed and leave it on the other side. The best used is the Comb harrow.

Disadvantages:

- All the seeds that are scattered do not have contact with the soil and hence 100% germination is not possible.
- to get the uniform plant population, enhanced seed rate is required.
- In this method, seeds cannot be placed in desired depth. Desired depth ensures perfect anchorage. Lodging (falling down) is common in broadcasting.

6.2.2. Dibbling

Dibbling is done by making a hole with equipment called dibble. After making a hole, the seed is inserted at a desired depth and the hole is then covered. Dibbling is performed on a plain surface and ridges and furrows or beds and channels. This method of sowing demands a suitable soil condition (Figure 6.2).

Figure 6.2. Sowing seeds through dibble.

Source: Image by Wikipedia.

Rice fallow cotton is sowed on a plain surface through dibbling. The seeds are dibbled at two-third from top and one-third at bottom of the ridge. Before starting the process of sowing, the furrows are opened, and fertilizers are applied and then the seeds are sown. The seeds and the fertilizers are not in contact with each other.

This method is adopted for the crops which needs wider space and medium to large size seeds. For Example: Sorghum, maize, sunflower, and cotton is dibbled on ridges and furrows.

The beds and channels and ridges and furrows come under line sowing. The plant occupies middle of the ridge while earthing up. For proper anchorage of the root system, earthing up is essential.

Advantages of sowing in line are:

- Uniform population;
- Better germination; and
- Reduced seed rate.

6.2.3. Sowing Behind the Plow

Sowing behind the plow is done manually and by mechanical means. The seeds are dropped in the furrows opened by the plow and after which the furrows are closed, and the new furrow is opened. Under this method, the seeds are sown at equidistance (Figure 6.3).

Figure 6.3. Manual method of sowing behind the plow.

Source: Image by Wikimedia Commons.

Since it is a manual method, therefore this is laborious and time-consuming process. Seeds like red gram, cowpea, and groundnut are sown behind the country plow. Major sown crop is ground out (Figure 6.4).

Figure 6.4. Mechanical method of sowing behind the plow.

Source: Image by Keverneland.

Other than manually, seeds are sown by mechanical means by Gorus-seed drill. A seed drill has a plow share and a hopper. Seeds are placed on hopper. There are different types of seed drill available, for example, simple Goru-Guntakas.

Advantages

- Under this method, the seeds are placed at a desired depth covered by iron planks.
- Except for very small seed, very large seeds can be sown, e.g., maize, sorghum, millets, sunflower, etc.

6.2.4. Drill Sowing (or) Drilling

The dropping of the seed in a definite depth covered with soil and compacted. Under this method, the sowing implements are used for placing the seeds into the soil. In this, there is both animal drawn Gorus and power operated implements are available (Figure 6.5).

Figure 6.5. Sowing seeds through drilling.

Source: Image by Agricultural Tourism.

This method ensures that seeds are drilled continuously or at regular intervals in rows. The depth of sowing is maintained and simultaneously fertilizers are applied. This method also provides the facility of sowing of inter crops. This takes lots of time, energy, and cost. However, it maintains uniform population per unit area. Seeds are placed at uniform depth, covered, and compacted.

6.2.5. Transplanting

This method of planting has two components.

- Nursery; and
- Transplanting.

In nursery, the seeds which are young are protected more effectively in a short period and a smaller area. The management is easy and economical.

Advantages:

- It ensures optimum plant population;
- Sowing of main field duration, which implies that the management in the main field is reduced; and
- Under transplanting, crop intensification is possible.

Disadvantages:

- Maintenance of nursery raising is expensive; and
- Transplanting is another laborious and expensive method.

The age of the seedlings is one-fourth of the total duration of the crop. Therefore, the total duration is 16 weeks, implies four-week period (one month) is under nursery beds.

However, nursery age is not very rigid. After the period of nursery is over, the seedlings are pulled and transplanted. This is mainly done through field preparation or optimum tilth on the main field. The seedlings are dibbled in lines an also in random.

Even after the nursery, the crops which are closer spaced are mostly raised by random method. For example, rice, and finger millet. During transplanting, desired spacing is required for vegetables.

After the transplanting, transplanting shock period comes where seedlings show no growth. This is because of the reason that the changes in environment change the relation between root and the soil. Therefore, the newly planted seedlings should adjust with the new environment. This environment ranges for 5 to 7 days depending upon the season, crop, variety, etc. At a higher temperature, leaves dry out and dehydration is possible.

The area required for nursery normally is one-tenth of the total area.

6.2.6. Hill Dropping of Seeds

The seeds under this method are dropped at fixed spacing and not in a continuous stream. Therefore, the spacing between plants to plant is constant. However, this is not the case in drills, where the seeds are dropped in continuous stream but the spacing between plant to plant in a row is not constant.

6.2.7. Check Row Planting

The check-row method of planting plants row-to-row and plant-to-plant and the distance between the seeds are uniform. This method ensures the plantation to be precisely along straight parallel furrows. The rows are in the two perpendicular directions. Check row planter is equipment used for planting the crops.

6.3. PRECAUTIONS TO BE TAKEN WHILE SOWING SEEDS

The most important part of the crop production is sowing seeds. It is important to focus on even the smallest details. While sowing the seeds, the following precautions must be taken:

The seeds are to be planted at a correct distance or intervals from each other. Because all the seeds planted must get their fair share of light, water, and nutrients for growth and development. to increase the yields of the farms, the seeds must be planted at equidistant from each other.

The depth of the seeds must be correct. Because if the seeds are roughly scattered on the top of the soil, it is most likely that the seeds will get blow away by the wind, or eaten by the animals or birds. In other case, if the seeds are sown very deep in the soil, the plant will not grow due to lack of air.

Therefore, it is always advisable that the seeds sown should be of the highest quality. They have to be free from the germ and disease.

6.4. GERMINATION

Germination is said to be the final event in the life of a seed. This stage in the life of the seed fulfills and completes the basic function of the seed. However, in modern agriculture, a seed has other functions as well. Those modern functions are the mechanisms by which the improvements are genetically engineered into the population of the plant and are then transferred from one crop generation to another.

Throughout the areas of adoption, they function as a convenient means of distributing plant populations. The functions discussed here are dependent on the germination process. There are cases when a seed is not able to germinate. This is due to the capacity of the seed that it has lost. This can neither transmit genetic improvements nor function in the distribution of desirable plant populations from one place to another.

Seeds are meant for propagating the crops and other species as well. The operations and activities involved in seed production are designed to maintain, protect, and/or enhance the propagative value of seed, i.e., capacity to germinate.

Seedsman should have a good understanding of the germination process and the vulnerabilities attached with the seeds. The resumption of the active growth of the embryonic axis in seed is called germination. This definition will be clear after a brief review of the events involved in the seed formation and development.

6.5. REQUIREMENTS FOR GERMINATION

The basic requirements for germination of seeds are moisture, a favorable temperature, and oxygen. These are explained in subsections.

6.5.1. Moisture

The requirement of moisture is needed for rehydration of the seed to the level that can increase respiratory activity. The breakdown of complex reserve materials such as starch, fats, and oils, and proteins into simple, mobile, and usable forms, and the synthesis of new materials for growth.

The moisture has to be in the liquid phase. Because seed cannot absorb enough water vapor to bring moisture content high enough to complete the germination process. The supply of water required for moisture content is generally supplied by media on which the seed is planted-soil, peat, blotters etc.

The seed that absorbs the water involves a special type of diffusion called as imbibition. Water or any other mobile material usually moves from a place where there is high concentration in volume. It flows in the low concentration by the process called as diffusion. This goes on until an equilibrium is achieved. This is done with an assumption that there are no barriers.

The water content in a seed at 10–13% moisture is very impure and hence not concentrated. The concentration is much lower than the water content in the moist blotter, damp peat, or even relatively "dry" soil. Therefore, the net movement of the water is from media (soil, peat, blotter, etc.) into the seeds.

The initial stages of the water absorption by a seed are the most important. They are usually the same whether it is a germinable seed, non-germinable seed, or a dead seed. As water begins to move into the seed, the cells of the seeds get rehydrated and start developing a pressure just as in the case of pressure in the tire.

As the internal pressure-hydrostatic pressure-increases, the rate of water absorption slows down. Eventually, a point comes where the hydrostatic pressure in the seed cancels out any remaining difference in diffusion pressure between water in the media and seed. After which equilibrium is achieved. The seed is said to be then fully 'imbibed.'

The seed moisture content required for germination corresponds to the "fully imbibed" condition roughly. There is different requirement for germination among various species.

It is mandatory to attain a moisture content of 30–40% by seed of a species of the grass family-grasses, cereals, small grains, etc. to germinate to proceed to the point where active growth is resumed.

There are varying requirements for moisture content in different seeds. The cotton, soybean, and peanut seed requires moisture content of 50–60% for germination. The varying requirement of moisture content is due to the differences in chemical composition of the seeds.

Earlier it has been mentioned that water absorption by seed involves a special type of diffusion called imbibition. Imbibition means the swelling of the imbibing material such as seed. This also implies the intake of great amount of water in relation to the initial volume and dry weight of the imbibing material.

The material that enables the swelling in the seed is the colloidal material which have the great affinity for water. It lets the seed to swell and they become hydrated.

However, the rate of water absorption by seeds is affected by several factors such as:

- The permeability of the seed covering to water;
- Initial moisture content of the seed;
- Temperature;
- The relative concentration or purity of water in the seed and the media;
- The extent of forces binding water to the media;
- The extent of contact of the seed with the water supply; and
- The chemical composition of the seed.

To some degree, the coat and covering of the seed generally restricts the water absorption by the seed, which is desirable in this case. There are some cases where the seeds seed covering is impermeable to water, and no water can be absorbed.

At the time of plantation, the condition is certainly undesirable. As the temperature increases, the rate of water absorption also increases, other factors being constant. In this, the initial moisture content is lower, the area of the seed in contact with water is larger. There is wider gap in the concentration of water in the media (high) and seed (low).

There are several field conditions that can reduce the availability of water to the level where the critical seed moisture content for germination cannot be attained. However, seed have a great capacity for absorbing the water. During the seed bed preparation, the lack of rain can result in a low supply of moisture in the soil through the planting depth.

Hence, germination can be slowed down considerably under such conditions. If the level of soil moisture is still lower, the seed might be able to absorb only that amount of water which is not enough for germination.

The seeds are "stored" in the soil at high moisture content and often at high temperatures, until the soil is resupplied with moisture by rain or irrigation. When the storage period is long, then the seed will deteriorate.

The soil is too dry for the seed to germinate as the water content is less in soil. However, they do increase in moisture content and swell. In Mississippi (US State), the soil temperature at planting depth range from 80°F during the night to over 100°F during the day. Soybeans do not live under such conditions.

Planting the seed in the fertilizer band results in reduction of the availability of the water to the seed. This is due to the reason that the fertilizer reduces the concentration of water in the band zone.

6.5.2. Oxygen

Oxygen is the second most requirements for the germination of the seed. The supply of oxygen is needed for the increase in the respiratory activity to provide energy to carry on the germination process. An atmosphere has an abundance of oxygen, it becomes limiting for germination only when its availability to the seed is blocked or impeded by some environmental factor or seed condition.

However, excessive moisture in the soil reduces the chances of the oxygen to get into the seed and the media displaces the oxygen in the pore spaces below the threshold level. There are many seeds that die because of water logging for more than 2 or 3 days.

There is dormancy on the seed due to covering or coat of some kinds of seed. This is because of the absorption of the oxygen. There are some kinds of seed like rice and some aquatic plants can germinate submerged in water. This implies a condition that severely limits the oxygen.

6.5.3. Favorable Temperature

There is a range of temperature that every seed grows in. The germination process in each kind of seed can start when that temperature is present. The germination process can proceed in reasonable time if it is not blocked by dormancy. For germination in various kinds of species, the classical work on seed germination has defined three cardinal points along with the temperature range for germination of a species.

The cardinal points are the minimum or base, optimum, and maximum or ceiling temperatures. There is difference in different kinds of seeds. The minimum or base temperature is the temperature below which the processes of germination do not proceed. The growth is not there to the point of visible growth of the embryonic axis within a "reasonable" period.

It is difficult to establish the minimum temperature for many seed kinds because of its dependence on time. However, the main effect of the minimum temperature is up to a specific point in germination process. When a 15-or 20-day period is allowed, the minimum temperature established in a 10-day germination period is usually higher.

The temperature below the minimum but above freezing is usually not lethal for the imbibed seed. It also does not cause death of the seed unless it is exposed for a long period. The interaction is seen between the seed rotting micro-organisms, or the seed that are susceptible to imbibitional chilling injury.

When the temperature is raised from the minimum to the optimum, the germination in the seed is usually rapid. The temperature above which the germination mechanism fails and there is no visible growth seen is called the ceiling or maximum temperature.

The maximum temperature is rather specific and easy to establish than the minimum temperature. Moreover, the temperature above the maximum is usually lethal. The imbibed seeds that are exposed to temperatures above the maximum die and rot within a few days.

The optimum germination is considered as the temperature at which the maximum percentage of seed germinates in the shortest period. The cardinal temperature for the process of germination and the interrelationships can be illustrated by using a line to represent the temperature range within which germination takes place.

Maximum temperature is much closer to the optimum temperature than to the minimum temperature. The rate of germination gets affected by reducing the temperature from the optimum to the minimum. As the temperature decreases from the optimum to the minimum, the germination process and early seedling growth become progressively slower.

Regardless of the time allowed, on increasing the temperature above the optimum, there will be decreases in both the rate and percentage of germination. The physiological condition of the seed strongly influences the effect of temperature on germination.

The seeds which are newly harvested are usually residually dormant and are specific in their requirement for germination. This is in relation to the dormancy condition the rice seeds which are newly harvested germinate best at 32°C, while newly harvested wheat seed do best at 16–18°C.

The time seeds lose their dormancy, the optimum temperature shifts to a slightly warmer level and the range of the temperature increases for the process of germination. For the requirement of the germination, the seeds become less specific. As the seeds deteriorate, the temperature again shifts.

There is more sensitiveness in the seeds with regards to temperature, the range of the temperature for germination shortens and there is a shift in the optimum temperature to a cooler level. There have been many researches done using the temperature as a constant. However, naturally, there is a rare chance of germination of seed under constant temperature.

The climate is variable as it is usually high in day and low at night in the most climatic areas. Therefore, it is to be noted that most of the species germinate in altering temperature than in constant temperature. The forage grass seed germinate best under such daily alternations of temperature like 15 to 25°C, 20 to 30°C, and 20 to 35°C.

There are several ways in which the temperature affects the germination process. The rate of germination usually increases from the minimum to the optimum level and sometimes above the optimum level. There is a pronounced effect of the temperature on the rate of respiration and the other factors included in the rate of water absorption.

There are detrimental effects of the temperature on the living things and the other biological matter as the temperature approaches the maximum.

The decayed and the abnormal seedlings increases in numbers and mostly all the seeds die in almost few days.

Therefore, it is necessary to understand the requirement of the temperature range within which the germination of seed will occur. This topic is of use not only in academics but in the real life too. It is therefore necessary to plant the crops in the most favorable temperature for better germination.

6.6. CROP STAND ESTABLISHMENT

For better crop production, good crop production is one of the most important features. The better crop establishment is identified as optimum plant population in fields. In a cropped field, the number of plants per unit area is called as plant population.

6.7. OPTIMUM PLANT POPULATION

The optimum plant production is the number of plants required to produce maximum output or biomass per unit area. If there is any increase beyond this stage, then there is either no increase or reduction in biomass.

6.7.1. Importance of Plant Population

Following are the importance of plant population:

- The yield of any crop depends on final plant population;
- The germination percentage and survival rate in the field are the factors on which the plant population depends;
- Under rain fed conditions, there will be depletion of the soil moisture in high plant population before maturity, while the low plant population will leave the soil moisture unutilized;
- It is necessary to use other growth factors like solar radiation efficiently when the soil moisture and nutrients are not limited;
- Due to wide spacing in the crops, an individual plant will yield more under low plant population;
- However, under high plant population, individual plant yield will be low due to narrow spacing leading to competition between plants; and
- Gradually, the yield per plant decreases as plant population per unit area is increased, but yield per unit area increases up

to certain level of population. That level of plant population is called as optimum population.

Therefore, to get the optimum plant population, maximum yield per unit area is necessary. So, the optimum plant population must be identified.

6.7.2. Factors Affecting Plant Population

1. Genetic Factors

i. Size of the Plant

- The spacing of the crop is decided by the volume occupied by the plant at the time of flowering;
- As compared to rice and wheat, plants of red gram, cotton, sugarcane etc. occupy larger volume of space in the field; and
- There are also chances that the varieties of the same crop differ in size of the plant.

ii. Elasticity of the Plant

- Elasticity of the plant is the variation in size of the plant between minimum size of the plant that can produce some economic yield to the maximum size of the plant that can reach under unlimited space and resources;
- In indeterminate plants, the optimum plant population range is high;
- The tillering and branching habit of the plants are the reasons of the elasticity in plants;
- Elasticity range is less for determinate plants like pearl millet, sorghum; and
- For indeterminate plants like cotton and red gram, more branches will be produced by the crop.

iii. Foraging Area or Soil Cover

- to get the maximum sunlight, the crop should cover the soil as early as possible;
- Higher the intercepted radiation more will be the dry matter produced; and
- Close spaced crops intercept more solar radiation than wide spaced crops.

iv. Dry Matter Partitioning

- The amount of solar radiation intercepted by the canopy which depends on plant density is the dry matter production; and
- The canopy expands more rapidly as the plant density increases, more radiation is intercepted and more dry matter is produced.

v. **Crop and Variety:** Depending on the crops and varieties, the plant population varies.

a. Environmental Factors

- Time of Sowing
  - When the seeds are sown at different time intervals, the crop is subjected to various weather conditions; and
  - Day length and temperature have much influence on the plant population among different weather factors. As low temperature retards growth, high plant population is required to cover the soil.

b. Rainfall/Irrigation

- In irrigation condition, the plant population is more and it is less under rain fed;
- Under more plant densities, more water is lost through transpiration; and
- Under adequate rainfall/irrigation, high plant population is recommended.

c. Fertilizer Application

- To fully utilize higher level of nutrients in the soil to realize higher yield, higher plant population is necessary;
- At optimum plant population, there are increases in the nutrient uptake; and
- In low fertility soil, high population leads to nutrient deficiency symptoms leading low yield.

d. Seed Rate

- Quantity of seed sown/unit area, viability, and establishment rate decides the plant population.

6.8. EQUIPMENT FOR SOWING SEEDS

6.8.1. Seed Drill

Seed drill is a machine that helps in placing the seeds in a continuous flow in furrows. This is placed at a uniform rate and controlled depth. It is done with or without the arrangement of covering them with soil.

- Function of Seed Drill

Following are the functions of seed drill:

- To carry the seeds;
- To open furrow in uniform depth;
- To meter the seeds;
- Placing the seed in furrows in an acceptable pattern; and
- Covering the seeds and compact the soil around the seed.
- Seed cum Fertilizer Drill

When seed drills are fitted with fertilizer dropping attachment, it is called as seed cum fertilizer drill. It distributes the fertilizer uniformly on the ground. This drill has a large seed box which is divided lengthwise into two compartments, one for seed, and another for fertilizers. Seed drill may be classified as:

- Bullock drawn; and
- Tractor drawn.
- Components of Seed Drill

A seed drill with mechanical seed metering device mainly consists of:

- **Frame:** Usually, this frame is made up of angle iron with suitable braces and brackets. This frame is so strong that it can withstand all types of loads in working condition.
- **Seed Box:** It is made of mild steel sheet or galvanized iron with a suitable cover. Sometimes there is clogging of seeds for which a small agitator is provided.
- **Covering Device:** After the seed has been placed in a furrow, this device helps in refilling the furrow. Covering of the seeds is usually done by patta, chains, drags, packers, rollers or press wheels, which are designed in various sizes and shapes.
- **Transport Wheel:** Two wheels are fitted on the main axle. There

are some seed drills that have got pneumatic wheels also. There is provision of wheels that have suitable attachments to transmit power to operate seed dropping mechanism.

6.8.2. Seed Metering Mechanism

The seed metering mechanism is the mechanism of the seed drill or fertilizer distributor that delivers the seeds or fertilizers from the hopper at selected rates. There are several types of seed metering mechanism:

- **Fluted Feed Type:** The fluted feed type is the seed metering device with adjustable fluted roller which collects and deliver the seeds into the seed tube. The mechanism of fluted feed type consists of a fluted wheel, feed roller, feed cut-off, and adjustable gate. This is used for different sizes of grains.

There are a feed roller and the feed cut-off device that is mounted a shaft that runs through the feed cups. Throughout its periphery, the roller carries grooves. The roller keeps rotating with the axle over which it is mounted It throws the grains out on the adjustable gate from where it falls into the seed tube.

The fluted rollers mounted at the bottom of the seed box receive the seeds into longitudinal grooves and pass on to the seed tube through the holes provided for this purpose. The length of the grooves exposed to the seed can be increased or decreased when the fluted wheel is shifted sideways. Hence, the amount of the seed is controlled.

- **Internal Double Run Type:** It is the seed metering device in which there are fine and coarse ribbed flanges in the feed wheel. Mounted on a spindle, there is a disc. This is placed in a casing fitted below the seed box. It has double faced wheel.

One face of the internal double-run type roller has a larger opening for larger seeds. While the other face has smaller opening for smaller seeds. At the bottom of the box, a gate is provided to cover the opening not in use. The rate of seeding can be varied by adjusting the speed of the spindle which carries disc.

- **Cup Feed Mechanism:** A mechanism that contains the cups and spoons on the periphery of a vertical rotating disc. It picks up the seeds from the hopper and delivers them to the seed tubes. It is supported with a seed hopper that contains two parts. The upper seed box is called as grain box and the lower one is called feed box.

The delivery mechanism of the seed consists of a spindle that carries a number of discs. And a ring of cups attached to the periphery of each disc. There is a spindle with a frame and an attachment which is called as seed barrel. With the rotation of the spindle, one disc with its set of cups also rotates which picks up some seeds and drops them into small hoppers.

There are two faces of the cups, one for larger seeds and the rate at which the seed barrel revolves. This type of mechanism is common in British seed drills.

- **Cell Feed Mechanism:** A mechanism in which the seeds are collected and delivered. This is done by a series of equally spaced cells on the periphery of a circular plate or wheel.
- **Brush Feed Mechanism:** Under this mechanism, a rotating brush regulates the flow of seed from the hopper. This mechanism is mostly used in the bullock drawn planters in the country that have Brush feed mechanism.
- **Auger Feed Mechanism:** It is a distributing mechanism. It consists of an auger that causes a substance to flow evenly in the field. The substance flows through an aperture at the base or on the side of the hopper. In many countries, fertilizer drills have got Auger feed mechanism.
- **Picker Wheel Mechanism:** A mechanism in which there is a vertical plate with radially projected arms. With the help of suitable jaws, these arms drop the large seeds into the furrows. For example, seeds of potato.
- **Star Wheel Mechanism:** A mechanism of feeding which consists of a toothed wheel. This wheel rotates in a horizontal plane. This conveys the fertilizer through a feed gate below the star wheel.
- **Furrow Openers:** For opening the furrow, there is a furrow opener are provided in a seed drill for opening a furrow. The seeds are conducted from the feed mechanism into the boot. This is the place where the seeds fall into the furrows.

6.9. TYPE OF FURROW OPENERS

There are different types of furrow openers and they are listed below.

6.9.1. Shovel Type

Shovel type furrow opener are widely used in seed drills. There are three types of shovels that are used. They are:

- Reversible shovel;
- Single point shovel; and
- Spear point shovel.

This type of furrow openers is best suited for stony or root infested fields. At the point where the boots are fitted, shovels are bolted to the flat iron shanks. The boots carry the end of the seed tubes. Springs are provided to prevent the shock loads which can occur due to obstructions. The construction of shovel type is easy, cheap, and repairable. Hence it is a very common to use with the seed drill.

6.9.2. Shoe Type

Shoe type furrow opener's best works in the trashy soils. It is the place where the seed beds are not smoothly prepared. They are prepared from the two flat pieces of steel welded together to form a cutting edge. Black cotton soil is the one which is best for its use. Shoe is made up of carbon steel having minimum carbon content of 0.5% with a minimum thickness of 4 mm.

6.9.3. Disc Type

They are of two types: (i) single disc type and (ii) double disc type.

6.9.3.1. Single Disc Type

There is a concave disc for opening the furrow. This type of furrow openers is found suitable where plant debris or trash mulches are used.

6.9.3.2. Double Disc Type

In double disc type, there are two flat discs for the furrow opener. These discs set at an angle to each other. Trashy land is the land that is mostly used by this method of furrow opening. Tractors that have the high speed use this form of furrow opening method. The furrow opener consists of:

- Tine;
- Shovel; and
- Seed tube.

Shovel is made of carbon steel having carbon content of 0.5% and a minimum thickness of 4.0 mm. Whereas, Seed tube is a tube which carries the seeds from the metering device to the boot. Through suitable boots and furrow openers, seed tubes are provided at the lowest lines. The minimum diameter of seed tube is 25 mm.

Boot is a part of the sowing machine that carries the seeds or fertilizers from the delivery tube to the furrow. It is bolted or welded to the tine.

6.10. PLANTER

The planter is used in the seeds that are large in size and that cannot be used by usual seed drills. A planter usually consists of:

- Hopper;
- Feed metering device;
- Knock out arrangement;
- Cut-off mechanism;
- Furrow opener; and
- Other accessories.

Each row of plater consists of a seed hopper. There is a seed metering device in a planter. There are various seed metering devices in a planter. However, the most common device consists of a rotating plate which is located at the bottom of seed hoppers. Some of the planters use vertical rotors as well as inclined rotors.

The horizontal seed plate is the most common which is used in the planter. There are suitable notches and holes in the horizontal seed plate. These holes are called as cell.

The type of notches on the plates decides the types of the plates, these are:

- Edge drop;
- Flat drop; and
- Hill drop.

The function of the edge drop is to carry the seed on the edge in the cell of the plate. Similarly, the flat drop carries the seed on the flat of the plate in the cell. In a hill drop, it can carry several seeds at a time by the cell that rounds the edge of the plate. Because it is large enough to do so.

The seeds which are required to be sown at equal interval between plants to plant are usually sown by planter.

6.10.1. Manual Rice Planter

Manual rice planter is used for transplanting the paddy seedlings. The unit consists of skid frame, movable tray, and seed picking fingers. On the inclined tray, mat type seedlings are placed. The seedlings are picked up by the fingers when it is picked downwards and are placed in the ground. During the return stroke, seedlings are left on the ground. Opener can control the plant to plant spacing. There can be five to six rows with comb type finger.

6.10.2. Rice Transplanter

The rice transplanters consist of:

- Air cooled gasoline engine;
- Main clutch;
- Running clutch;
- Planting clutch;
- Seeding table;
- Float;
- Star wheel;
- Accelerator lever;
- Ground wheel;
- Handle; and
- Four bar linkage mechanism.

6.10.3. Seedlings

The most technical and difficult work for this type of transplanter is the growth of the seedlings. They are usually grown in special seedling trays which are known as mat seedling.

Growing of seedlings for this transplanter is most technical and difficult work. Seedlings are grown in special seedling trays. It is called mat seedling. Mat seedlings are grown by some standard procedure in controlled environment in a nursery.

6.10.4. Operation

Special trays are used to raise the seedlings as mat seedlings. The mat seedlings are placed in a slanting position on a seeding table. The running clutch and the planting clutch are operated just after the engine starts. There is four bar linkage mechanism which catches 3 or 4 seedlings at a time and separate them from the mat and fix in the puddled soil.

There is a float that supports the machine on the water while working on the field. The movement of the transplanter is supported by two iron wheels made of iron. While the machine is functioning, there is marker that demarcates the transplanting width.

The power is transferred to the main clutch and then it is divided into two routes. This power then goes to the planting clutch and the other goes to running clutch. The mechanism will not work unless the planting clutch is operated.

6.10.5. Soil Preparation

While starting the agricultural process, preparing soil is the first step., it is concerned with loosening of the top layer of the soil which is known as topsoil. It is the most essential part of the agricultural process as the plants absorbs water and food from the soil. Hence, to produce a healthy crop, soil preparation plays an important role.

The soil is prepared by the following methods:

- Plowing;
- Leveling; and
- Manuring.

6.11. CONCLUSION

Crop geometry is the spacing between the individual crops. Crop geometry is same as plant geometry. In crop geometry, the yield of the crop has influence on light interception, rooting pattern and moisture extraction pattern. The plant geometry can always be changed by changing inter and intra-row spacing or the planting pattern.

The different crop geometry shapes are square plantation, rectangular plantation, random plantation, paired row plantation etc. There are different methods of sowing seeds and transplanting them. The first method being the broadcasting followed by drilling, dibbling, transplanting, planting, and putting seeds behind the plow.

The planting pattern decides the yield of the crop. Therefore, it is necessary to plant the seeds in a way that results in high yields of the crops. There is numerous equipment that helps in faster and safer sowing of seeds. That equipment is discussed in this chapter.

REFERENCES

1. Agrinfobank, (2013). Crop Geometry | Agriculture Information Bank. [online] Available at: http://agrinfobank.blogspot.com/2013/11/crop-geometry.html (accessed on 2 April 2020).
2. Delouche, J., (n.d.). Seed Germination. [eBook] Available at: https://ir.library.msstate.edu/bitstream/handle/11668/14130/1990%261992-06-DeloucheSeedGermination.pdf?sequence=1&isAllowed=y (accessed on 2 April 2020).
3. n.d. Seeds-Seed Rate-Sowing Methods-Germination-Crop Stand Establishment-Planting Geometry. [eBook] Available at: http://eagri.org/eagri50/AGRO101/lec11.pdf (accessed on 2 April 2020).
4. n.d. Sowing and its Equipment. [eBook] Available at: http://www.hillagric.ac.in/edu/coa/agengg/lecture/243/Lecture%2013%20Sowing%20equipment.pdf (accessed on 2 April 2020).
5. Toppr-guides, (n.d.). Soil Preparation and Sowing of Seeds: Methods with Videos and Examples. [online] Available at: https://www.toppr.com/guides/biology/crop-production-and-management/soil-preparation/ (accessed on 2 April 2020).

CHAPTER 7

Weather and Weather Forecasting

CONTENTS

Weather is a combination of temperature, wind direction and speed, amount, and type of precipitation, sunshine hours, etc. It is defined for a short period for a particular place. Weather forecast is the prediction of the climate condition beforehand. This chapter discusses the issues related to the climate and weather information dissemination.

The first half of the chapter defines the weather and weather forecast. Going forward, the chapter consists of some topics such as, weather forecasting for agriculture, its usefulness, and characteristics of weather forecast. The chapter ends with the topics, weather, and the crops, weather, and the farmers.

7.1. INTRODUCTION

In agricultural production, the most important role is played by weather conditions. The weather condition has an influence on the crop growth, development, and yields; on the incidence of pests and diseases; on water needs; and on fertilizer requirements. These influences are the result of the differences in nutrient mobilization.

Since the water stresses and the timeliness and effectiveness of preventive measures and cultural operations with crops. The weather aberration renders the soil and crop to erode. This is the physical damage done to the soil and crop.

The quality of the crop depends on its journey from the field to the storage and the transportation to different areas of the country. It is the movement of the crop that decides the quality of the agricultural produce. Along with this, the weather conditions play an important role in the quality matters of the crop.

Bad weather conditions may affect the crop production during its transportation from the field to the consumer area. Therefore, it can be said that there is no aspect of crop culture that is immune to the impact of weather. The optimal growth of the crop depends on the weather factors. There exists a role in the incidence and the spread of pests and diseases.

Various crops are affected by the weather induced stress and the affliction by pests and diseases. The crop affected is among different varieties within the same crop, and different growth stages within the same crop variety.

On a climatological basis, there is a spatial variation in an area at a given time, temporal variations at a given place, and year-to-year variations for a given place and time. to produce the crop, the weather over a short period or year to year fluctuations at a particular place over the selected time interval has to be considered.

At a given unit of time, the percentage departures of extreme values from a mean or median value, called the coefficient of variability. These are the measure of variability of the parameter. The degree of the variability of a given weather parameter is greater in the shorter time unit. There are intensities of the variations that differ among the range of weather factors.

In time and space, the most variable of all the parameters is the rainfall over the short periods. The short period inter-annual variability for the rainfall is large. This implies that the variability needs to be expressed in terms of the percentage probability of realizing a given amount of rain. In other words, at a given level of probability, the minimum assured rainfall amounts need to be specified.

to achieve the optimum productivity at a given location, the crops, and the cropping practices must be such that while their cardinal phased weather requirements match the temporal march of the relevant weather element(s), endemic periods of pests.

While the diseases and hazardous weather is avoided. The routine and the processed short-period climatic data play a vital role in the strategic planning of crops and cropping practices.

Weather forecast is becoming more important task for efficient management of farm inputs. The accuracy of forecast for a long period is a great challenge, but short and medium range weather forecasts in recent year were improved to a great extent due to modern technology. Agromet advisory bulletin is the best tool for transferring agro techniques and location specific recommendation of SAU"s to the end users. Int.J.Curr.Microbiol. App.Sci (2018) 7(2): 2644-2653 Agromet Advisory Bulletin - A Weather Smart Agriculture Technology option for Adaptation and Mitigation of Changing Climate P.R. Jaybhaye*, N.H. Deore and P.B. Shinde https://doi.org/10.20546/ijcmas.2018.702.321

7.2. WEATHER EVENTS

There are various types of weather events that exists on a year-to-year basis. This is due to the careful agronomic planning on a microscale to suit experience in local-climate crops. The deviations from the normal weather conditions occurs in almost all years, areas, and seasons with higher frequencies.

The most common is the delay in the start of the crop season due to rainfall vagaries in the case of rain fed crops. The other being the temperature (as observed in the tropics, temperate zones and subtropics), or persistence of end-of-the season rains in the case of irrigated crops (Figure 7.1).

Figure 7.1. Weather events.

Source: Image by Pixabay.

There are other phenomena that deviates from the normal features in the temporal march of various weather elements. However, the effect of the deviations of the weather on crops builds up slowly but are often widespread enough to destabilize national agricultural production.

7.3. WEATHER FORECASTING FOR AGRICULTURE

Weather forecasting is defined as the prediction of the state of the atmosphere for a given location. This is done after applying the principles of physics and supplemented by a variety of statistical and empirical techniques and the technology.

Along with the prediction of the atmospheric phenomena themselves, the weather forecasting predicts the changes on Earth's surface caused by atmospheric conditions (Cahir, 2013).

The forecasting of the weather conditions is important to the human beings for preserving the life and property, to save crops and to tell us what to expect in our atmospheric environment. This is the reason humans have started to predict the weather informally for millennia and since the 19th century for a variety of public and private uses.

There were some authors who believed that there is an economic value behind the weather forecasts and tried to estimate it through different methods (Craft, 2010). The public uses this forecast for various severe weather alerts and advisories to protect lives and minimize losses. This range to the air and marine traffics which are both very sensitive to the weather.

There are some utility companies such as electricity, gas which rely on the weather forecasts to anticipate demand which can be strongly affected by the weather.

Weather forecasting is also useful in the forests' management for preventing and controlling wildfires and for predicting conditions for the development of harmful insects. It has private usage as well. In case of private companies, to avoid large losses and increasing their profits, the companies pay for the weather forecast according to their needs.

One such example is that the supermarkets may change their stocks on their shelves in different weather conditions. They may also change the stock and the reason may be the in anticipation of different consumer spending habits in different weather conditions. Similarly, the military weather forecaster's present weather conditions to the war fighter community.

7.4. SUMMARY ADAPTATION DESCRIPTION

The summary adaptation description is considered as a common practice. In this practice the crop economics, infrastructure, and water resources allow. In this practice, the soil type is being dictated, the water availability to a crop is to be monitored, and the potentially supplemented is done through the irrigation scheduling.

This practice is mostly done for the potatoes and soft fruits (Suŝnik et al., 2006). This practice is also done on cereals and some salad crops (Wilks and Wolfe, 1998). The coordination in the integration of the weather forecasting into the irrigation scheduling system and decision making is said to be less common.

However, there is a wider adaptation that is being done on the skillful weather forecasts, over the time horizon of the hours to the months ahead. It also offers the potential for the more efficiently uses the scarce of water resources in the agriculture applications. They are also used for the economic benefits (Wilks and Wolfe, 1998; Freebairn and Zillman, 2002).

The term skillful is defined as the forecast method is more accurate than simply assuming future weather conditions will remain the same (persistence) or that weather conditions will simply be average for the time of day or year (climatology). The skillful weather forecasts that are modern are based on the numerical weather prediction approach, involving computer-based modeling from a set of initial atmospheric and environmental conditions.

The accuracy of the weather forecast can be very sensitive to the reality of the specified initial conditions. This highlights the importance of the high-quality observational data and its geographical coverage, both horizontally and vertically. The predictability of the weather condition varies with time. In case of irrigation scheduling, there is no need of having daily forecast granularity. For example, a week is going to be wetter or drier than the average for the next four weeks can also assist with the irrigation decision-making. The forecast of the precipitation amount (P) and evapotranspiration (ET) are relevant in the case of weather forecasts which relate to water usage in agriculture. ET depends upon forecasts of air temperature, humidity, surface solar radiation intensity, and wind speed (Figure 7.2).

Figure 7.2. Wind and weather instrument.

Source: Image by Public Domain Pictures.

The difference between the precipitation and ET forms the basis for the decision-making with regard to irrigation scheduling. The forecasts of cumulative (P-E) over days or longer will normally represent the most relevant averaging period.

The evaporation is dependent on the vegetation/crop type, and the generic (P-E) forecast estimates can be made more locally relevant. This is done through the incorporation of this information. The crop type and its stage of development are taken into account.

In case where the supply of the soil moisture is limited, there the actual ET may not match the facts of the potential Evo transpiration. This simply implies that the forecasts of P-E may be needed to get it adjusted to the account for this. This can be handled as a so called "Soil Moisture Deflect."

This process is done either through the on-farm soil moisture measurement process or through the computer modeling of the soil water balance (Smith et al., 2006; Hough, 2003; Hough and Jones, 1997).

7.5. CURRENT STATUS AND UPTAKE

The in-field monitoring of soil moisture is being coupled by the licensed alert system driven by the software or by the advisory service that is, being provided by a consultant and it is the most current operational irrigation scheduling systems. These systems are mostly considered as the current or recent in-field conditions, rather than forecast conditions.

Few integrated decision-support systems do exist where a weather forecast is fully embedded in the irrigation scheduling system. There is modest number of studies in the literature describing the potential benefits (Venäläinen et al., 2005; Cai et al., 2011).

However, the attention is paid to the forecast weather conditions, and is generally done by a separate online forecast link embedded within an irrigation scheduling software system. It can be done through the ad-hoc grower monitoring that is done by one or more (often free) independent channels of weather forecast information (e.g., TV, radio, website, mobile phone app etc.).

The significant water usage efficiencies arising from the weather forecast information are likely to require the widespread integration of skillful and the fit for the purpose forecasts. The forecast is done in the existing popular irrigation scheduling software systems. There must be a trust placed by farmers in weather forecast information more generally (Hu et al., 2006; Wang and Cai, 2009).

There is a requirement in the communication between the user needs and the potential service providers of the agriculture (Sivakumar, 2006)., the idea of the weather prediction is the choice of the sample of the state of the fluid at a given time and uses the equations of fluid dynamics and thermodynamics. The thermodynamics helps in estimating the state of the fluid at some time in the future.

There are different forecasting techniques used for the purpose, which are discussed in subsections.

7.5.1. Persistence

The simplest way of forecasting the weather is the persistence technique. This technique usually relies upon the today for making the forecast for tomorrow. This method is useful in the short run as well as the long run forecast. However, there is a need of a strong presence of a stagnant weather pattern (Sauter, 2005).

7.5.2. Use of a Barometer

Since the late 19th century, the measurement of the barometric pressure and the pressure tendency (the change of pressure over time) has been used in forecasting. The larger the change in pressure, the larger is the change in the weather (The old farmer's Almanac, 2013).

- If the drop of the pressure is rapid and a low-pressure system is approaching, then there is a greater chance of rain.
- The rise in the rapid pressure are associated with improving weather conditions.

The condition of the sky is as important as the pressure tendency. This parameter is helpful in forecasting weather mainly in the mountainous areas. The use of the sky cover method to predict weather over the centuries has led to various weather lore.

For instance, predicting the rain through the observation of the thickening of cloud cover or the invasion of a higher cloud deck. The high thin cirrostratus clouds at night can cause halos around the moon that indicates an approach of a warm front and its associated rain.

7.5.3. Use of Forecast Models

Earlier, the human forecaster was responsible for forecasting the weather. The scenario is changing now; today model based forecasting is followed. This forecasting is done on the basis of the available observations.

Now, the humans are only confined to choosing a model on the basis of some parameters such as model biases and performance. To reduce the errors in forecasting, use of a consensus of forecast model and the ensemble members of various models is done. However, the forecasters are required to interpret the model data into the forecasting. Then, compare the prediction of the model with the actual observations. If forecaster feels necessary, they can change the forecast.

7.5.4. Analog Technique

This technique is a complex technique for making a forecast. This is based on a previous weather event expected to the same in an upcoming event. The main difficulty is that the same event is rare to happen in the future.

There are less chances of a perfect analog for an event in the future. However, this method is useful in forecasting the rainfall and the forecasting of future precipitation and distribution (Ben Daoud et al., 2011).

Since 1959, when Vanguard 2" was launched, to forecast the weather observations, satellites, and weather forecasting satellites were being used. They usually use the links to list some things that are observed from the space. There are total two types of satellite orbit; polar orbiting and geostationary and they both are used in meteorology (Figure 7.3).

Figure 7.3. Symbols for weather forecast.

Source: Image by Free SVG.

When the traditional data sources are missing, data from the weather satellites are used. There is an advantage of using the data from the satellite that it has the global coverage as compared with any other technique being used. The other techniques are still struggling with lower accuracy and the resolution.

7.5.5. Suitability and Target Use

Weather forecast service has various advantages that are available across the agriculture. The farming practice that is irrigation-fed is one of the obvious methods of agriculture. The forecasting of the weather conditions provides advanced indication of the output of on-farm renewable energy systems (e.g., wind, solar and anaerobic digestion systems).

This can also help in the optimum scheduling of a wide range of on-farm activities for avoiding wasteful use of resources and to protect the environment. The weather forecast is also helpful for the animals of the farm. It is evident in the case of chickens, that the weather and the humidity that has to be adjusted according to their survival needs. This enables the optimum storage conditions for crops.

The prevention of the agricultural practice can be achieved by the good and reliable forecasts. This manages the extreme events of the weather. Nowadays, the weather forecast models operate at higher spatial resolution than ever.

The biases between the output of a model and the local conditions experienced on-farm can mean that it is alright to post-process the raw forecasts to more faithfully reflect the actual farm climate.

The availability of the on-farm weather station measurements supports the post-processing to calibrate the adjustment process. With respect to the on-farm soil moisture measurements which are few and far between official meteorological monitoring networks, there is an opportunity for sharing the data between the growers and forecast providers.

According to world meteorological organization (WMO) (2010), the weather forecasting is divided into the various groups. The shorter the range, the highest the predictability. The forecast then done is the most important and the suitable for agriculture. The groups are as follows:

- **Now-Casting (NC):** This is the description of the current weather variables and description of forecast weather variables for 0 to 2 hours. This enables the production of a relatively complete set of variables such as air temperature and relative humidity (RH), wind speed and direction, solar radiation, precipitation amount and type, cloud amount and type.

- **Very Short-Range Forecast (VSRF):** VSRF is the weather forecasting method that is used for the explanation of weather variables that can range up to 12 hours. There is a relatively complete set of variables that can be produced.
- **Short-Range Forecast (SRF):** SRF is used to describe the weather variables that is between 12 hours to 72 hours. In this also, a relatively complete set of variables can be produced.
- **Medium Range Forecast (MRF):** MRF is the description that ranges from 3 days to 10 days. Through this method, relatively complete set of variables can be produced.
- **Long Range Forecast (LRF):** From 12–30 days up to 2 years. There is some restriction based on this method and that is some fundamental variables, temperature, and precipitation. On the other hand, the elements of agricultural weather forecasts vary from place to place and from season to season.

However, they should refer to all weather elements, which affect farm planning and operations, and they ideally would include (WMO, 2010):

- Sky coverage by clouds;
- Precipitation;
- Temperature (maximum, minimum, and dew point);
- Relative humidity;
- Wind Speed and direction;
- Extreme events (heat and cold waves fog, frost, hail, thunderstorms, wind squalls and gales, low pressure areas, different intensities of depressions, cyclones, tornados);
- Bright hours of sunshine;
- Solar radiation;
- Dew;
- Leaf wetness;
- Pan evaporation;
- Soil moisture stress conditions and supplementary irrigation for rain fed crops;

- Advice for irrigation timing and quantity in terms of pan evaporation;
- In some of the cases, the specific information about the evolution of meteorological variables into the canopy layer; and
- In some specific cases, micro-climate inside crops.

7.5.6. Design and Management Issues

There is big improvement in the development of weather forecast reliability and accuracy. This is all because of the numerical weather prediction models (NWP). However, the predictability is currently limited to about 3 weeks for temperature and 2 weeks for precipitation and solar radiation. The precipitation forecasts over a month are encouraging (Calanca et al., 2011; WMO, 2010).

This is the reason that most of the research is devoted to the improvement in such models, other empirical models, and methods derived from experience (rules of thumb). At some point of time in the future, humans will not be required to make the forecast as the development of new powerful tools and models is constantly evolving (Met Office, 2009).

However, this will involve greater efforts and will fetch lot of investment to carry out the research. Also, this will need greater technological performances. For example, faster super computers and greater coverage and the targeting of observations, monitoring tools that use satellites, networks of automatic weather stations, radars, and lightning detection systems.

The oldest technique is the Nowcasting for very short-range forecasting. It dates back to the 1860s. However, in 1980s, the method to forecast is gaining accuracy and reliability with the development of the NWP models and is expected to improve with time to deliver high reliable data for end users (Met Office, 2013; Wilson, 2006).

7.6. USEFULNESS OF WEATHER FORECASTS

The occurrence of bad weather condition is beyond the control of human. However, the bad weather conditions can be adapted or mitigated if the forecast is received on time. In rural areas, people depend on the rule of thumb for anticipating the local weather and timing of agricultural operations in light of expected weather.

According to Basu (1953), there is no scientific basis for anticipation of weather in many of the popular proverbs and folklore. In a recent study, the conclusions of Banerjee et al. (2003) matched with that of Basu (1953). The proverbs show that the rural people and the farmers were keen to know the likely weather situations for crop operations from time immemorial.

There are some agronomic strategies that are available to cope with changing weather. For example, if there is delay in the crop season, it can be countered by using a short-duration varieties or crops and thicker sowings.

However, once the crop season starts, the resources and the technology get committed. The only way left is to adopt the crop-cultural practices. This minimizes the effects of mid-seasonal hazardous weather phenomena, while relying on advance notice of their occurrence.

For example, the use of irrigation or lighting trash fires can prevent the effects of frosts. Therefore, the medium range weather forecast with a validity period that enables the farmers to arrange all the appropriate cultural operations is useful. This enables the farmers to take the advantages of the forecasted weather.

There has been a rapid advancement in the information technology. This technology spread to the rural areas has provided farmers with meeting their demands. The farmers are benefitted from the timely and accurate weather forecast.

7.7. WEATHER FORECASTS FOR AGRICULTURE: ESSENTIAL REQUIREMENTS

The forecast that calls for a late start to the crop season should result in the departures from normal agronomic practices at the field level. This involves a high cost and requires a considerable amount of time. This is the reason that the pre-season forecasts must have a validity period of at least 10 days and more than a week.

The forecast of the hazardous weather, pests, diseases, and the like cannot be implemented instantaneously. The mid-season forecast should be communicated within 5 days in advance or in 3 days (the least).

The dissemination of the weather forecast must be communicated to the users as soon as possible with lowest time lag following their formulation. There are some methods that include high cost such as pre-season agronomic corrections, control operations against pests and diseases, supplementary irrigation, and the scheduling of early harvests (Figure 7.4).

Figure 7.4. Weather prediction for agricultural farm.

Source: Image by Pxhere.

Thus, the weather forecast should be accurate along with the timeliness of the forecast. Small area weather forecast is easy as there are well-organized weather systems and the desired areal delineation of forecasts can be realized. However, in other cases, the weather forecast will be applicable, and it must be unambiguously stated.

7.8. SOME UNIQUE ASPECTS OF AGRICULTURAL WEATHER FORECASTS

There is difference in the aspects of the forecast of the agriculture and the synoptic weather forecast. In synoptic forecast, the withdrawal of the monsoon is the related to changes in wind circulation patterns in the upper atmosphere. This also relates to the associated changes in precipitable water content of air in the lower layers.

For the crops to grow, there is need to prepare the field for sowing and the sowing of the crop with adequate availability of the seed-zone soil moisture requires a good amount of rain. There are agriculturally significant rains (ASRs) (Venkataraman, 2001), that enables the commencement of the cropping season and contribute to the water needs of the crops. The rain that do not contribute to the root-zone soil moisture of standing crops are ineffective.

For the agricultural purposes, the start and the end of the ASR are important. The significant rain for the crops can be received as thundershowers or it may be delayed. There may also be difference in the climatological dates of withdrawal of the monsoon and the end of ASRs in a region.

On a climatological basis, the use of dependable precipitation (DP) at various probability percentage levels and potential evapotranspiration (PET) has been suggested for delineation of the start and end of a crop growth period (WMO, 1967, 1973; Venkataraman, 2002) and have been used in many regions.

The methods are different in terms of the time units employed. The probability level chosen for the DP and the fraction of PET used as a measure of adequacy of crop rainfall. Venkataraman (2001) suggested that weekly period be used. The commencement and end of the ASRs be the point at which DP is at 50% probability level begins to exceed PET and becomes less than 50% of PET, respectively. On the basis of the considerations of the level of the evaporative power of air (EPA), the rainfall amount required to overcome the evaporative barrier, and phased moisture needs of crop demands.

The commencement and the end of the ASRs can easily be delineated when the rainfall probability data for weeks and the monthly values of PET are available. However, there is a need of clear weather for the sowing operation to continue, it must be preceded by the soil moisture storage.

Therefore, the forecasting of the clear weather and the wet spell is crucial. The forecasts of dry spell following the wet spell are required for starting the disease control measures. There are some areas where before the rain comes, thunderstorm takes place. Once the rain starts, it persists without any stop in between.

In the above case, for preparing the land, the agronomic strategy must be adopted to utilize pre-season rains. The dry sowing must be done in anticipation of rain to come in the next few days. Land preparation must be started when the thunderstorm has arrived as a sign that the rain will also come in few days or hours.

It is a prudent idea to sow when there is forecasting of the rain in the coming days. The dry-sown seeds will be baked without the rain. Thus, the rainy season forecast is very crucial in such areas. In regions, where there is frost, it can pose a serious threat to the crops. Frost usually comes when there is 0°C.

7.9. CHARACTERISTICS OF WEATHER FORECASTS

As per a deterministic definition of the weather forecast, "weather forecast describes the anticipated meteorological conditions for a specified place (or area) and period."

Another definition states that, "weather forecast is an expression of probability of a particular future state of the atmospheric system in a given point or territory."

According to the above definitions, it is clear that weather forecast is the declaration in advance the likelihood of the occurrence of future weather events at a given period at a given place. When the weather forecasters are aware of the farmer's requirement and they know how to utilize the forecasted weather. The positive effect of the weather forecasts in agriculture is maximized.

Response of the varieties of the crop to a weather phenomenon is the degree and not the type. The type and the intensity of the weather phenomenon that causes setback to the crops vary among the crops. Crop weather factors are the factors which mean that the crops and the cropping pattern vary across areas, even within the same season.

The first step that is being used by the forecasters is to provide the weather warning that is required by the farmers, so as to have a preparation of crop guides according to forecasters. Forecaster should also provide the time of occurrence and the duration that is required for the development of the phrases. This development occurs from sowing to harvest, that is major crops in the regions of the forecasts interest.

Forecaster should also specify the types of weather phenomena that are used for the weather warnings. Forecast should be issued in the different phrases of the crops.

These guides can be used by forecasters to prepare calendars of agricultural weather warnings with a breakdown by periods and regions. For the agrometeorological work, this crop guide must be useful. These guides contain the normal values of important weather elements in the crop season. These are used on a national level work of agriculture.

This guide must also be made available to the farming community so that the farmers knows in advance the weather conditions before starting any agricultural practice.

7.10. CONSIDERATIONS RELATED TO AGRICULTURAL WEATHER FORECASTS

The elements of the weather forecast refer to all the weather elements that immediately affect farm planning or operations. However, the elements vary from place to place and from season to season. Normally a weather forecast includes the following parameters.

Weather forecast includes the following parameters.

- Amount and type of cloud cover;
- Wind speed and direction;
- Rainfall and snow;
- Relative humidity;
- Minimum, maximum, and dewpoint temperatures; and
- Extreme events, such as fog, different intensities of depressions, frost, tornadoes, hail, thunderstorms, heat waves, cold waves, wind squalls and gales, low-pressure areas, cyclones.

An agricultural weather forecast should also contain the following information:

- Bright hours of sunshine;
- Solar radiation;
- Soil moisture, supplementary irrigation and stress conditions are for rain fed crops;
- Dew;
- Leaf wetness;
- Pan evaporation;
- Advice for irrigation timing and quantity in terms of pan evaporation;

- Microclimate inside crops in specific cases; and
- Specific information about the evolution of meteorological variables into the canopy layer in some specific cases.

7.11. WEATHER AND CROPS

Solar radiation, air temperature, humidity, and precipitation are the key atmospheric variables that impact the crops. The variability of it in day to day of the landscape can be described as weather. The weather extremes at critical periods of a crop's development can have adverse impact on the productivity and yields.

The climate can be described as the long-term average temperature and humidity and the total solar radiation and precipitation over a crop's growing season. For any given region, the climate decides the realized yields.

7.11.1. Crop Response to Weather Variables

It is difficult to describe the weather variables and its response to different climatic conditions. In case of one variable being limited, the effects of solar radiation or precipitation do not greatly affect the crop. In another scenario, where there is no limiting variable, the crop responds to the variable that is farthest from the optimum for that variable.

The description of the physiological response of the crop at the field level poses additional uncertainty in predicting crop yield. There is a greater uncertainty in the prediction of the exact response of the crop to the weather. This is based on the normal weather conditions.

There is a rare chance of the normal year to occur. However, there is a variability of the weather normal across years and within the year. Some period being the wetter/drier, hotter/cooler, sunnier/cloudier than normal.

7.11.2. Temperature

Temperature is the main variable that decides whether a crop will grow or not. Along with the precipitation and solar radiation, it tells that how well a crop will grow and how fast it will develop. There are four temperature thresholds for determining the growth of a crop and these are given as: the absolute minimum, the optimum minimum, the optimum maximum, and the absolute maximum.

The absolute minimum and maximum temperatures describe the coldest and the hottest temperatures at which a crop will grow. The best performing crop is described by the temperatures between the optimum minimum and maximum.

7.11.3. Temperature Stress

Heat and cold both are responsible for the crop stress. Heat stress is experienced in the summers and the cold stress is experienced in the spring and fall season. Most of the agronomic crops are not affected by the cold stress. But in summers, the heat stress causes severe damage to the crop.

When the temperature is greater than the optimum maximum of the crop, the crop experiences the heat. A heat stress unit (HSU) is the number of degrees the maximum daily air temperature is above the heat stress threshold times the number of days.

Plants are also affected by the heat stress because temperature increases, respiratory reaction rates speed up, using more of the photosynthetic compounds manufactured in a day. If the soil is not able to meet the additional water requirement, the stress caused by heat is compounded by an added water stress.

7.11.4. Soil Temperature

In the autumn season, the soil temperature is defined when ammonium nitrogen fertilizer may be applied without excessive nitrification occurring during the autumn and winter. At a depth of 4 inches (10 cm) below 50°F (10°C), the rate of nitrification is reduced with the soil temperature. The process however becomes negligible only when soil temperatures are below 32°F (0°C).

The estimation of soil temperature can be done by computing the average of the mean air temperature for the preceding 7 days. In the autumn season, the estimates so found tend to overestimate soil temperatures by 1–2°F (0.5–1.0°C). The error in the estimate creates a conservative estimate of the soil temperature.

7.11.5. Growing Season Length

The period between the last spring frost and the first fall frost is the growing season length. Generally, frost occurs when the minimum temperature is less than or equal to 32°F (0°C). This is the reason that most of the crops are planted after the major risk of the frost has passed.

However, there are circumstances when the perennial and the annual crops gets damaged due to the late frosts or very late frosts. In the spring season, the frost damages the crops. The frost usually freezes at a temperature less than 30°F (–1°C). In many countries, the first frost occurs before or on October and November. The normal growing season generally less than 170 days and more than 185 days.

7.12. WEATHER AND FARMERS

There had been great efforts in the efforts to provide information regarding the weather forecast and the climate information to the farmers. However, these arguments that suggest these improvements should not be taken as panacea.

As per Stern and Easterling (1999), In terms of forecast information in particular, 'climate and weather forecasts in their current form are often ill-suited for direct use in decision-making and decision-making is often ill-suited for the use of weather and climate information.'

The farmers from the developing countries or the farmers of those countries whose markets are well established are mainly benefitted from the weather and climate forecasts. The farmers of the developing countries are benefitted due to their high vulnerability. There is no doubt that the greatest benefit may go to the farmers who have the means to take the most advantage of the technology.

The smallholder farmers in most of the developing countries, these would be those farmers who could apply productivity-enhancing technologies such as improved seeds, fertilizers, and labor. In other case, there is limited demand especially for the farmer-oriented climatic information, since the options available to farmers are limited (Stern and Easterling, 1999).

The prerequisites for potential benefits of meteorological and climatological forecasts are pointed out by Hansen (2002a), if it is to be realized by the farmers. These forecasts are meant to address the need of the farmers that is 'real and perceived.' The weather forecast and the climate need to have the relevant component of the climate or weather system at an appropriate farming scale.

Most importantly, Hanson has pointed the benefit that also depends on the existence of decision options for the farmer. The decision options are sensitive to the particular incremental information that the forecast provide.

The information must also be compatible with the farmer's goals. There is need to interpret the information appropriately the relevant forecast. The interpretation is needed to be made with sufficient lead-time to make an impact on their decisions.

The prediction of the relevant components of weather or climate variability is needed at relevant period at an appropriate scale. The prediction is supposed to be accurate and reliable and must be in the form that a farmer can apply it in the decision making. Yet there are many agencies who may not understand the process as the appropriate climate and weather information is needed.

7.13. REAL-TIME WEATHER AND CLIMATE INFORMATION 'PRODUCTS' FOR FARMERS

Motha and Stefanski (2006) have provided many examples of weather and climate information as data-rich products from many government organizations. These organizations provide fundamental ongoing information for farmers, farmer organizations, agricultural consultants, industry, and other agriculturally focused government or semi-government organizations.

The government of many countries has successfully applied weather and climate information in many of its operational applications that are disseminated to farmers and other agricultural users. The routine and special crop-weather assessments provide critical information to the decision makers who formulate crop production forecasts and trade policy. This is achieved when integrated with economic analyses and information.

The examples of the tactical products include the long running "Weekly Weather and Crop Bulletin." The other consists of routine and special crop weather assessments, and regional weather networks. Other specific products include publications and reports, a drought monitor, crop calendars, as well as international crop and weather information of value to those with interest in commodity prices and similar issues.

The farmers who are in the drought-prone regions can access climate, drought, and crop prospects using information provided on specialized information sites (Figure 7.5).

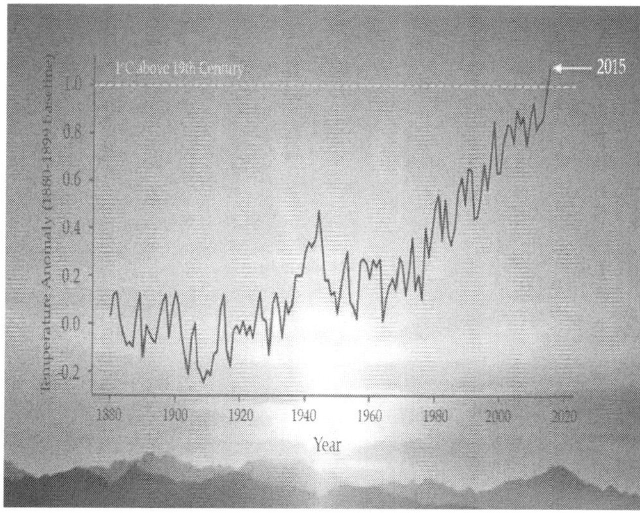

Figure 7.5. Graphing global temperature trends.

Source: Image by JPL NASA.

7.14. ENGAGING WITH FARMERS: THE VALUE OF A PARTICIPATIVE APPROACH

The farmers who have successfully understood the complex climate and weather information, it becomes important for them to participate in the development of appropriate response strategies to climate and weather information. They have to decide that which decisions relates to the climate forecast information that suits them.

In this respect, the farmers are suspicious of the forecast system if they do not understand or have some ownership of the scientific methods used to develop it. Especially, if the forecast is conflicting with their local traditional indicators (Patt and Gwata, 2002).

The value of the forecasts to farmers depends on their accuracy. The farmers also depend on the management options available to the farmer to take advantage of forecasts (Nicholls, 1991, 2000). This is true that the value of the climate and weather forecasting to the farmer may never have been demonstrated to the farming community by the institution which develops and promotes the forecast information.

The decision-making of the farmers depends on the cost and benefits of different decision options. This determines the minimum level of skill for forecast information to the farmers (Katz and Murphy, 1987; Hansen, 2002a b; Gadgil et al., 2002).

Sonka et al. (1987) reinforced the fact that to identify those areas where tactical changes can be made. This was done to take advantage of predicted (probabilistic) above-average rainfall and to reduce losses in predicted (probabilistic) below-average situations.

Or else, the climate and the weather forecast, they must not have any value, unless the key management decisions that have been identified through the close interaction with the farmers. With this, the farmer's key management decisions are capable of changing by the incorporation of the climate and weather forecasts information's (Nicholls, 1991, 2000; Hammer et al., 2001).

In terms of seasonal climate forecast systems, there are some agricultural and climatological agencies that are issuing the seasonal climate forecast that is on the basis of operational. There also few chances that can result in the failure by these agencies to get engage in a suitable participative approach. There are also chances to develop a comprehensive profile of the users and the key decisions that may be manageable to the forecast system.

Following this failure, it has created a considerable gap in the between the information that is likely to be useful to farmers and that provided and disseminated by these agencies. It is suggested that the vulnerability to major climate phenomena such as El Nino events and their dependence on national and international aid will be less, if farmers use climate forecasts to make better decisions.

There are also some common problems that are leading to confusion among the farmers by using climate forecasts and information, which is the failure in differentiating between its 'value' or 'impact 'and the 'skill' of a forecast (Murphy, 1993, 1994; Amissah Arthur, 2000; Hammer et al., 1996; Hammer, 2000; Weiss et al., 2000; Hammer et al., 2001; Patt and Gwata, 2002). Cash and Buizer (2005) helps in emphasizing the effective climate and weather information systems that should 'ground the collaborative process of problem definition' in farmer's perspective.

This instance would lead to the shift the focus towards promotion of broad, farmer-driven risk management objectives. Rather than necessarily advancing the uptake of particular forecasting or information technologies by the meteorological institution.

Hansen (2002b) emphasized on the point as a core and urgent requirement in bridging the institutional and cultural gap that exists between meteorological and agricultural support institutions.

The meteorological institutions are used in regarding to provide the information system and it also forecasts as a standalone products. People who are considered as a agriculturally oriented professionals, farmers takes these systems as a process which include assessing the information system and climate forecast which can be used to increase the farm productivity.

to overcome this problem, while ensuring that the objectives are farmer driven. It has been suggested that a knowledge-action system needs to be evaluated appropriate to the achievement of the farmers' ultimate goals (e.g., more effective risk management).

Instead of the goals that are used for the meteorological community, for example, more or better understanding and using the information and forecasts, there is a goal that is used for improving the content, distribution, format that are used order to increase the use and impact (Hansen, 2002b; Cash and Buizer, 2005).

In a wide varying region in many parts of the countries, the study of seasonal climate forecasting applications for farmers. It has been noted that while farmers were interested in the receiving seasonal rainfall forecasts that provided the probability of receiving a total rainfall amount in millimeter over their farming season. The farmers were rather interested in receiving the forecasts that were more relevant to their actual decisions.

This can include the timing of the commencement of the wet season, whether there would be interruptions in rains, and whether the period used as output from the forecast system was relevant. Once this additional detail is provided, especially the use of participative approaches with farmers.

They regard the climate forecasts and weather forecasts as capable of providing useful input into their key management decisions (Gadgil et al., 2002; Everingham et al., 2002; Ingram et al., 2002).

There is urgent need globally for meteorologists to increase their communication activities with farmers as per Sivakumar (2002). Cash and Buizer (2005) suggested that the designing of full end to end system for starting the developmental process the climate and weather forecaster must visit the farmers and listen to their perspective, their problems, and their needs.

Everingham et al. (2002) also emphasized that the focused conversation with farmers reveal that climate and weather information are needed as a part of larger suite of information that helps them manage a broad array of risks. The key to success is the initiating conversations with lead innovators within the farming community.

This fosters the broader participation of the other farmers and improved connection between scientists and farmers (Cash and Buizer, 2005). The participative approaches with the farmers may reveal that they may have an adaptive management strategy in place. This is independent of the weather and climate information or forecasts.

One of the characteristics of managing and farming cropping and grazing systems is the requirement for a certain lead-time between deciding on a course of action and realizing its results (Carberry et al., 2000; Carter et al., 2000). The importance of getting the 'big (strategic) decisions' right is stressed such as land purchase, machinery investment, and resource improvement.

According to Meinke and Stone (2005), the farmers are better if they solve their problems roughly rather than to find an extreme well way of solving it. In farming practices, it is better to be rough in the decisions than to be wrong (P. Hayman, South Australian Research, and Development Institute, personal communication).

7.15. INTEGRATED SYSTEMS APPROACHES TO APPLICATION OF CLIMATE AND WEATHER INFORMATION

The overall benefit of the application and integration of the information and forecasting with farm management strategies is improved with the application of seasonal climate forecasting.

This includes similar systems in the application whole farming system scale and across industry value chains. This approach is helpful in many areas that have the potential to benefit industries.

Taking an example for sugar farming, that could also have application in a general sense to other industries, the strategies include:

- Using better scarce water resources improves on-farm profitability, increasing the use of water efficiency, and enabling higher production with consequent minimal movement of nutrients and pesticides off-farm.

- There must be improved planning for wet weather harvest disruption. The early season supply and better scheduling of milling operations leads to more effective use of resources (for example, milling capacity, haulage capacity, haulage equipment, shipping, together with enhanced on-farm profitability).
- There is enhanced industry competitiveness through more effective forward selling of the commodity. This is based on enhanced knowledge of the amount of supply and improved efficiency of commodity shipments.

The probability distribution of a large number of simulated yields and gross margins can be produced and incorporated into risk assessment tools. This determines the value of the integrated climate/crop modeling efforts.

There is exploration of climate influences such as ENSO on extreme outcomes which is allowed by the large number of simulations using the modeling approach. This approach is difficult with purely historical series that are typically short (Sivakumar, 2002; Podesta et al., 2002; Meinke and Stone, 2005).

The decisions that are benefitted from more integrated and targeted forecasts are made at a range of temporal and spatial scales. This includes the tactical decisions of the scheduling of planting or harvest operations to policy decisions with land use allocation (e.g., grazing systems vs. cropping systems).

It is useful to think about all the aspects of climate/weather/farming systems research and farmer decision-making from a systems perspective.

Manton et al. (2000) proposed the end-to-end concept of The RES AGRICOLA project. If farmers are going to get the benefit, it focuses on three discipline groups that need to interact closely. These fundamentally important discipline groups are:

- Climate sciences.
- Agricultural systems science (including economics).
- Rural sociology (Meinke and Stone, 2005).

When the system employs farmers, other decision makers, and scientists (climate/meteorology/biophysical) across various disciplines. It ensures that the issues that are addressed are relevant to the farmer (Meinke et al., 2001). The facilitation of improved pay-offs for farmers are significant.

To use such a framework which is sustained requires the institutional commitment and favorable policies Hansen (2002a).

7.16. LINKING CLIMATE AND METEOROLOGICAL INFORMATION TO FARMER ACTION

An argument given by Cash and Buizer (2005) says that the linking climate and meteorological knowledge to farmer action will be more effective if it is perceived by farmers to be simultaneously salient, credible, and legitimate. These terms are defined as follows:

7.16.1. Salience

The perceived relevance of climate and weather information is the salience. It generally means that does the system provide the information to the farmers they need and at the time, they need and can use the information?

7.16.2. Credibility

Credibility addresses the perceived technical quality of the information. It generally means that does the provide information that is perceived to be valid, accurate, tested, or more generally, at least as likely as alternative views to be "true."

7.16.3. Legitimacy

It is concerned with the system that the system has the interests of the farmers in mind. It is not only for pushing the agendas and interests of other actors. Patt and Gwata (2002) has proposed six factors limiting farmers' use of climate forecasts together with suggestions for corrective action.

Moreover, there may be some other important factor that significantly inhibit uptake of climate information and forecasts by farmers. The obstacles are identified by Selvaraju et al. (2004), and it has been classified as:

- The inherent underlying issues related to the forecast system itself;
- The nature of smallholder cropping or farming systems for which the forecast systems were to be employed;
- The inherent underlying aspects that is associated with the socio-economic characteristics of the farmers;

- Related to issues associated with the information delivery systems; and
- Related to external policy and external institutional factors.

There are chances of considerable alteration in the presence of seasonal climate forecasts. In modeled analyses, the farm production decrease in less favorable years was shown as less substantial than in any an analysis of actual actions of resource poor farmers.

It is always suggested that in case of resource-poor farmers with few production options, the most common occurrence is decrease in the area planted in potentially poor rainfall seasons, increase in production volatility. This occurs when the forecasts are readily and widely available.

When the crop investment is expected to be costly, it is assumed that there would be reduced production by small farmers in years. The important thing is that how farmers react to the climate forecast. This will determine whether aspects of increased uncertainty with regards to farming production will result in avoided costs or missed opportunities (Hill et al., 2001; Phillips et al., 2002).

There is a need of further investigation as a rational response when there is decrease in the total farm production in the years when the climate forecast suggests increased chance of below-normal rainfall (Phillips et al., 2002). In a potentially poor season, instead of planting risky crops, it is always advisable to opt for Off-farm employment and income in potentially poor years.

Also, a poor farmer cannot turn short-term gain into the savings for the poor production years in the future (Phillips et al., 2002). Hammer, 2000 used a simple example of tactical management of row configuration in a cotton crop and demonstrated the basis for effective application of seasonal climate forecasting on-farm.

7.17. DISSEMINATION OF CLIMATE AND WEATHER FORECASTS AND INFORMATION TO FARMERS

The dissemination of the climate and weather forecast information to the farmers depends on the effectiveness of the strongly on the systems that distribute the information, the channels of distribution, farmers' modes of understanding and judgment about information sources.

The climate and weather forecast information is complex that the effectiveness of the information is affected. The ways in which the information is presented is also important (Stern and Easterling, 1999).

There are some farmers who may have difficulty in understanding the forecast information. This may lead them to obtain and translate this information into a deterministic version erroneously. According to some rural belief systems, it does not incorporate the idea of "chance or luck." It may be the result of the previous actions.

The policy makers may communicate the information to the farmers deterministically that tells the nature of the climate forecast information. However, communication of the forecast information to the farmers is the most important barrier.

The producers of climate forecasts have not given much attention to the communication process as an integral component of an effective climate information system (Changnon, 2000; Patt and Gwata, 2002; Podesta et al., 2002).

These arguments emphasize the need to reinforce the compelling need for climate forecast information to be prepared and disseminated in a very timely manner. This should be done to link with key decision times in a farmer's planning schedule. There is only one time that the farmers apply the management information that is aided by weather or climate forecast information.

If the climate forecast information is issued too late for effective management systems completely neglects the efforts of putting the forecast information at first place. Patt and Gwata (2002) has stressed on the fact that it is only through the repetitive communication that developers of climate forecast systems and farmers learn about each other's methods. This way they come to know who makes the decisions, what decisions they make, and through what channels information arrives.

Agromet advisory bulletin (AAB) format invented by Jaybhaye et al 2018 and adopted by IMD , disseminating the weather forecast and weather based advisory under the Gramin Krishi Mausam Seva (GKMS) found beneficial. P.R. Jaybhaye*, N.H. Deore and P.B. Shinde. 2018.Agromet Advisory Bulletin - A Weather Smart Agriculture Technology option for Adaptation and Mitigation of Changing Climate . Int.J.Curr.Microbiol.App. Sci (2018) 7(2): 2644-2653. https://doi.org/10.20546/ijcmas.2018.702.321

Fig.3 AAB pictures format suitable to multimedia and end users (Format 2)

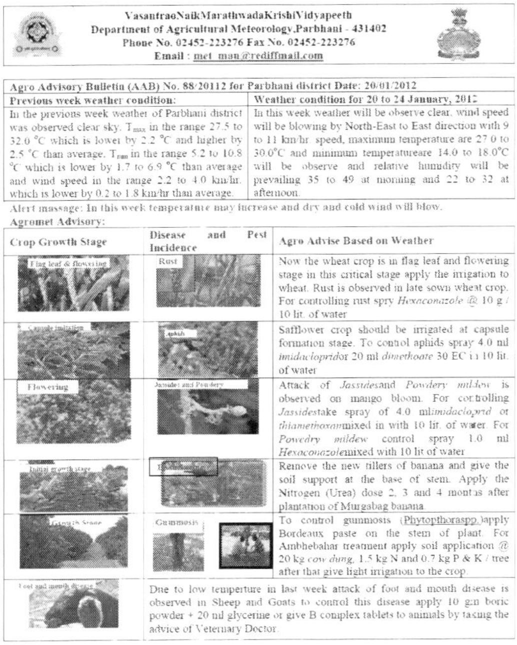

VasantraoNaikMarathwadaKrishiVidyapeeth
Department of Agricultural Meteorology,Parbhani - 431402
Phone No. 02452-223276 Fax No. 02452-223276
Email : met_man@rediffmail.com

Agro Advisory Bulletin (AAB) No. 88/20112 for Parbhani district Date: 20/01/2012	
Previous week weather condition:	**Weather condition for 20 to 24 January, 2012**
In the previous week weather of Parbhani district was observed clear sky. T_{max} in the range 27.5 to 32.0 °C which is lower by 2.2 °C and higher by 2.5 °C than average. T_{min} in the range 5.2 to 10.8 °C which is lower by 1.7 to 6.9 °C than average and wind speed in the range 2.2 to 4.0 km/hr. which is lower by 0.2 to 1.8 km/hr than average.	In this week weather will be observe clear, wind speed will be blowing by North-East to East direction with 9 to 11 km/hr. speed, maximum temperature are 27.0 to 30.0°C and minimum temperatureare 14.0 to 18.0°C will be observe and relative humidity will be prevailing 35 to 49 at morning and 22 to 32 at afternoon.

Alert massage: In this week temperature may increase and dry and cold wind will blow.

Agromet Advisory:

Crop Growth Stage	Disease and Pest Incidence	Agro Advise Based on Weather
Flag leaf & flowering	Rust	Now the wheat crop is in flag leaf and flowering stage in this critical stage apply the irrigation to wheat. Rust is observed in late sown wheat crop. For controlling rust spry *Hexaconazole* @ 10 g / 10 lit. of water
Capsule initiation	Aphids	Safflower crop should be irrigated at capsule formation stage. To control aphids spray 4.0 ml *imidacloprid*or 20 ml *dimethoate* 30 EC in 10 lit. of water
Flowering	Jassides and Powdery	Attack of *Jassides*and *Powdery mildew* is observed on mango bloom. For controlling *Jassides*take spray of 4.0 ml*imidacloprid* or *thiamethoxam*mixed in with 10 lit. of water. For *Powedry mildew* control spray 1.0 ml *Hexaconazole*mixed with 10 lit of water
Initial growth stage		Remove the new tillers of banana and give the soil support at the base of stem. Apply the Nitrogen (Urea) dose 2, 3 and 4 months after plantation of Murgabag banana.
Growth Stage	Gummosis	To control gummosis (Phytopthoraspp.)apply Bordeaux paste on the stem of plant. For Ambhebahar treatment apply soil application @ 20 kg *cow dung*, 1.5 kg N and 0.7 kg P & K / tree after that give light irrigation to the crop.
Foot and mouth disease	Due to low temperture in last week attack of foot and mouth disease is observed in Sheep and Goats to control this disease apply 10 gm boric powder + 20 ml glycerine or give B complex tablets to animals by taking the advice of Veternary Doctor.	

Agromet advisory bulletin (AAB) prepared by GKMS centre based on current weather, forecasted weather and real time field crop condition have a useful tool for enhancing the production and income through which improving agro-economic status of farmers. Pralhad Jaybhaye*, P.B. Shinde and R. Balasubramanian . 2018. ADAPTATION OF AGROMET ADVISORY BULLETIN (AAB) FOR IMPROVING AGRO-ECONOMIC STATUS OF FARMERS. Inno. Farm., 3(1): 24-30

7.18. FARMERS' NEEDS FOR DECISION-SYSTEMS RELATED TO CLIMATE CHANGE

Managing the future farming risks which is associated with long term climate change is regarded as managing low frequency components of climate variability. The same quantitative approaches that were developed for the planning issues in farming of climate variability can be applied to the farming issues that are likely to occur under the change in climate (Meinke and Stone, 2005).

Although there are various complex issues associated with the climate change such as changes in land cover and changes in runoff. Many farmers are of the view that they will cope with the adverse effects of the climate change.

However, according to some studies, it is believed that farmers will continue to produce the same commodities on the same land using the same management tools (Abler and Shortle, 2000; Abler et al., 2000; Rogers and McCarty, 2000).

7.19. CONCLUSIONS

Weather and the climate forecast system in its current form have the potential to provide improved farm management and profitability worldwide. The forecast information may not be suitable in all the decisions of the farmers. Some farmers may not be able to use the forecast information directly in their decision making.

to overcome this problem, it is important that the farm decisions to be identified that are sensitive to the information which forecasts provide. This should also be compatible with the farmer's goals. Therefore, the climatic and weather forecast information has to be tailored in the way that suits the key decisions of the farmers.

The climate and weather forecast information have no value if it is not able to turn around the management decisions. The emphasis on the participatory approaches with farmers, there can be considerable gains that can be made in linking the climate and weather forecast information with the decision of the farmer.

REFERENCES

1. Balasubramanian, A., (2017). Elements of Climate and Weather. [online] ResearchGate. Available at: https://www.researchgate.net/publication/319181402_elements_of_climate_and_weather (accessed on 2 April 2020).
2. Hollinger, S., & Angel, J., (n.d.). Weather and Crops. [eBook] Available at: http://extension.cropsciences.illinois.edu/handbook/pdfs/chapter01.pdf (accessed on 2 April 2020).
3. Stone, R., & Meinke, H., (2006). Weather, Climate, and Farmers: an Overview. [eBook] Available at: https://rmets.onlinelibrary.wiley.com/doi/pdf/10.1017/S1350482706002519 (accessed on 2 April 2020).
4. Weather and Climate Forecasts for Agriculture, (n.d.). [eBook] Available at: http://www.wamis.org/agm/gamp/GAMP_Chap05.pdf (accessed on 2 April 2020).
5. Weather Forecasting for Agriculture, (n.d.). [eBook] Available at: http://web1.adas.co.uk/cranfield/Documents/Weather%20forecasting.pdf (accessed on 2 April 2020).

CHAPTER 8

Climate Change and Impact on Agriculture

CONTENTS

This chapter named as 'Climate Change and Impact on Agriculture.' This chapter describes the climate change in brief manner. In the first half of the chapter, the climate change and adaptation and also the causes and effects of the climate change is described.

The warming, climate change and impacts have been mentioned in the chapter. Climate change in context with agriculture has been described in the second half of the chapter. This part of the chapter mentions the impact of climate change on agriculture and furthermore, it provides insights about how agriculture helps in fighting climate change.

8.1. INTRODUCTION

In today's world, climate change can be cause by a number of different activities. For example, the dramatic changes in the temperature can result in more flood, can cause droughts or intense rain, as well as it can cause more frequent and severe heat waves.

Oceans as well as glaciers have also experienced some of the changes such as oceans are warming and becoming more acidic, glacier are melting and there is increase in the sea levels. These changes will likely present challenges to our society and environment because these changes frequently occur in future decades.

Considering the past century, the human activities have released an abundance of carbon dioxide and some other greenhouse gases into the atmosphere. There are a number of gases which come from burning the fossil fuels to produce the energy (Figure 8.1).

Figure 8.1. A massive glacier.

Source: Image by Flickr.

Greenhouse gases are said to be a blanket which is present around the Earth, and this traps the energy in the atmosphere which eventually causes warming. This is referred to as greenhouse effect and it is natural as well as important to support the life on earth. However, along with the building up of the greenhouse gases, the climatic changes can further result in dangerous effects to the health of human as well as ecosystems.

Stable climate has been adapted by the people since the last ice age which had ended several thousand years ago. There are changes which can take place due to the warmer climate and this can affect our water supplies, agriculture, power, and transportation systems, the natural environment and also, our health and safety.

Among the climatic changes, there are some which are unavoidable and there is nothing which can be done about it. Let's take an example. In the atmosphere, the carbon dioxide can stay for around a century, so that the Earth will continue to warm in the coming future.

Over the last century, in the world, the global warming has really taken effect. This is not usual that there is rapid increase in the average surface temperature of the Earth over the past century primarily because of the greenhouse gases that is released as people burn the fossil fuels.

The global warming is caused because of the enhancing the greenhouse gases emissions and the build-up in the environment of the Earth.

The gases have an influence on the atmosphere and these gases are water vapor, carbon dioxide, nitrogen oxide, and methane. Approximately 30% if the incoming sunlight is reflected to the space which is done due to the bight surfaces such as clouds and ice. From the other 70%, most of that is absorbed by the land as well as ocean and the rest are absorbed by the atmosphere.

The solar energy which is absorbed tends to heat the planet. This absorption and radiation by the atmosphere are supposed to provide benefits to the life living on the Earth. Considering the time today, the atmosphere contains much more greenhouse gas molecules.

The infrared energy is emitted by the surface and this ends up being absorbed by the atmosphere. The atmosphere of the Earth is becoming to be a more efficient greenhouse and this is happening by the increasing concentration of the greenhouse gases.

Throughout the history of Earth, the climate has now cooled and warmed for a number of different reasons. In the history of the planet, there is rapid warming which is considered to be not normal.

There are some factors which have an effect on the climate such as volcanic eruptions and changes in the amount of the solar energy which are natural. Climate can even change if in case there is a change in the amount of solar energy reaching to the Earth.

The climate can get affected by the volcano eruptions. The reason is that when it erupts then it spews out than just lava and ash. There are tiny particles, which are released, made of Sulfur dioxide (SO_2) into the atmosphere.

Such particles get into the stratosphere and reflect the solar radiation out to space. Snow as well as ice has a great amount of effect on the climate. Along with the melting of the snow and ice, the Earth tend to warms up, then less energy is reflected further causing even more warming.

There are a number of various ways which plants, animals, and some other life on the planet Earth which can affect the climate. From these ways, there are some which can produce the greenhouse gases and these gases tend to trap earth and further causing the global warming through the greenhouse effect.

From the atmosphere, the carbon dioxide is taken out by the plants as they make their food by using photosynthesis. During the night time, the carbon dioxide is released by the plants back into the atmosphere.

When the farm animals such as cattle and sheep tend to digest their food then the methane is formed. Cars and trucks tend to release carbon dioxide further affecting the climate, this happens when the fossil fuels are burned so as to power them.

The carbon dioxide is released into the atmosphere, when there is occurrence of wildfires. Nevertheless, if a forest of the same size grows again then same amount of carbon, which was added to the atmosphere due to the fire, will be removed.

As per the prediction of the scientists in the past world, there are some effect which would result while the changes in the world is occurring such as loss of sea ice, accelerated sea level rise and much more intense heat waves. The scientists are much confident that the global temperature will continue to rise for upcoming decades, majorly because of the greenhouse gases which are produced by the human activities.

According to the Intergovernmental Panel on Climate Change (IPCC), the extent to which climate change tends to affect the individual regions will vary over time and having the ability of different societal and environmental systems mitigate or adapt to the changes.

Since 1880, this decade was considered to be the warmest. According to the National Oceanic and Atmospheric Administration, the year 2010 and the year 2005 has been the warmest years.

During the 21st century, the earth could warm by an additional of 7.2°F if in case, the reduction in the emissions from the burning fossil fuels fails. There can be much great effects of the rising temperature on the climatic patterns of the Earth and all the living beings living on Earth.

There are industrial activities on which the modern civilization depends, these activities have raised atmospheric carbon dioxide from 280 parts per million to 379 parts per million, considering the last 150 year.

8.2. CLIMATE CHANGE AND ADAPTATION

The rise in the burning of fossil fuels and land use changes have emitted a number of gases and so, is continuously emitting the same, further increasing the quantities of the greenhouse gases into the atmosphere of the Earth.

Such greenhouse gases include CO_2 that is Carbon Dioxide, CH_4 that is Methane and N_2O that is Nitrogen Dioxide. The rise in these gases have caused an increase in the amount of heat that is received from the sun withheld in the atmosphere of the Earth, heat that will normally get radiated back into the space.

Such increase in the heat has resulted in the greenhouse effect, furthermore leading to the climate change. The major characteristics of the climate change are:

- The increases in the average global temperature (that is global warming),
- Melting of ice caps and glaciers and reduction in the snow cover,
- Changes in the cloud cover and precipitation specifically over land, and
- Increasing ocean temperatures and ocean acidity, because of the absorption of the heat and carbon dioxide from the atmosphere.

There was a fourth assessment report presented by the IPCC that is Intergovernmental Panel Climate Change (IPCC, 2007) and this dispelled a number of uncertainties about the climate change. Now, the warming of the climate is unequivocal. Now, it is clear that most of the time, the global warming is because of the emissions of greenhouse gases which are man-made (mostly that is Carbon dioxide).

Over the last century, there was increase in the atmospheric concentrations of carbon dioxide from a pre-industrial value of 278 parts per million to 379 parts per million in the year 2005. There was rise in the average global temperature that is by 0.74°C. According to scientists, this is considered to be the largest as well as fastest warming trend which they have been able to discern in the history of the Earth.

In the last 25 years, there was an increase in the rate of warming which has particularly taken place and 11 of the 12 warmest years (on records) have occurred in the last 12 years.

According to the IPCC report, there are the detailed projections which have been provided for the 21st century and these represent that the global warming will continue as well as accelerate. There were best estimations done which in turn indicates that the Earth could warm by 3°C by the year 2100. The Earth will continue to warm up despite that there is reduction in the greenhouse gas emissions by the countries. There were predictions by 2100 and this range from a minimum of 1.8°C to as much as 4°C increases in the average global temperatures. There is a variable climate around the human being for centuries to which they have been adapting.

The decision of people can be influenced by the worldwide local climate variability having consequences for their social, economic, political, and personal conditions and so, have effects on their lives as well as livelihoods.

The effects which are because of the climate change imply that the local climate variability that the people have in the previous time experienced and have adapted to is changing at relatively great speed.

8.3. CAUSES AND EFFECTS OF CLIMATE CHANGE

Climate change can be defined as the seasonal changes that have occurred for a long period of time in the world. Such climate patterns play an important role to shape the natural ecosystems and the human economies and cultures which depend on them.

Nowadays, the climates of Earth are changing in a rapid manner and much differently. Along with this, the climate changes in the Earth are affecting the lives physically, emotionally, and also psychologically.

Let's take an example, the changes in the climate can affect the way in which the human, plants, and animals live like the production of food and health risks. As the climate of the Earth is dynamic and so, is always changing through a natural cycle, now, therefore, the world is worried about the changes which are occurring today and also, its speed which is increasing as well as seriously affecting the lives of human.

Considering the whole world, it can be said that the people are taking action so as to prevent the climate change. For example, in the year 2007, the scientists from IPCC that is International Panel on Climate Change predicted that the oceans are warming up and the glaciers are melting which is because of the global warming and climate change which eventually can cause rise in the sea levels up to 7–23 inches by the year 2100. Therefore, there are some islands and some small countries which will be flooded by the sea water and can even get dispersed very soon (Figure 8.2).

Figure 8.2. Icon for climate change mitigation.

Source: Image by Wikimedia Commons.

It is now important to take action as soon as possible so as to try to resolve this problem or to prevent the much more serious occurrence of the climate changes.

Natural causes and human activities are said to be the two main causes of the climate changes. The natural causes have influenced the climates of Earth such as volcanic eruptions, ocean current, the orbital changes of Earth, and the solar variations. There is a cooling effect on the earth and that is caused by the eruptions of volcanoes.

The large volumes of Sulfur dioxide (SO_2), dust, water vapor and ash are thrown out into the atmosphere which happens when the volcano erupts. The climatic pattern is also affected for years due to the volcanic eruptions despite the fact that the eruptions occur in a few days. The SO_2 gas will reach the upper level of the atmosphere.

There are tiny particles, dusts, and ashes which block the incoming sunrays and this will result in the cooling at the atmosphere. This is due to the cooling of atmosphere of the Earth that occurs because of the bouncing of the sunlight to space.

Along with this, ocean current is considered to be one of the natural causes which affect the climate changes. In the climate system, the ocean is the major component. Around the 71% of the Earth is covered with the oceans and it absorbs around twice as much as the sun's radiation as the atmosphere or the land surface. The winds push horizontally which is against the sea surface and further drive the ocean current patterns.

Additionally, the oceans tend to play an important role of the concentration of the carbon dioxide. There can be changes in the ocean circulation which affect the climate by the movement of the carbon dioxide into or out of the atmosphere. Furthermore, there is another because which can affect the climate change and that is the orbital's changes of the Earth.

Every year, the earth tends to make one full orbit around the sun. If there is no tilt then there will no such experience of seasons. The changes that occur in the tilt of the Earth, then this can affect the severity of the seasons.

Let's take an example, there is more tilt then this will mean that the human will experience warmer summers and colder winters and if in case, there is less tilt then this means that the human will experience cooler summers and milder winters.

There is one more main cause which results in the climate change and that is human activities. From the 19th century, the Industrial revolution has experienced the large-scale use of the fossil fuels for the industrial activities.

Therefore, this has created a number of jobs for the people. And a number of people have moved from rural areas to the urban areas. A large number of vegetation areas were cleared to build houses or factories for the industries.

The natural resources are extensively used for construction, industries, transport, and consumption. All of these activities have provided their contribution in rising the greenhouse gases such as methane and nitrous oxide in the atmosphere.

Industries factories and houses have been built by human and vehicles are being used which used many fossil fuels such as coal, natural gases and oil. So, climate changes tend to be the result of this use of the fossil fuel.

Global climate change is caused because of the fertilizer which is used for the agricultures. The microbial action in the soil result in the release of the nitrous oxide. Along with this, the emission of methane as well as carbon dioxide which will also affect the climate changes. Moreover, the increase in the agriculture is also the result of the increasing human populations in the world.

There are a number of causes and all the causes are having a great impact on the climate change in our Earth. There are three main effects on the climate changes in the environment. First of all, in the world, the climate change tends to affect the human health. There are many researchers which have represented that the climate change can result in the human health and further, it produces diseases.

The second one is that the climate changes will also tend to affect the biodiversity. The biodiversity is very important considering the plants and animals to maintain their habitat as well as ecosystem. The climate change has also affected the loss of biodiversity.

Let's take an example for the same. There can be changes in the water sources, the food chain can even destroy and the medicine sources can also change. Along with this, there can be effects on the marine ecosystem also which is because of an increase in the sea temperature as well as ocean acidification. This in turn increases the vulnerability of fragile ecosystems such as coral reefs.

Furthermore, the human minds and behavior can also get affected because of the climate changes. Considering the human behaviors, the climate changes will affect the consumption, energy use and population growth. Additionally, the climate changes can also affect the psychosocial as well as mental health which includes stress, guilty, anxiety to adapt and cope with the weathers which are continuously changing.

There will be increase in the social disorder which will caused by the climate changes such as post traumatic disorder, suicide, abuse, anxiety, stress, and violence. Therefore, this can be said that the climate changes will be providing a great impact on the psychological lives of human.

8.4. WARMING, CLIMATE CHANGES AND IMPACTS

The scientists have made a great stride while predicting the amount of temperatures which tend to change that can be expected for different amounts of future greenhouse gas emissions. Further, they have understood the way in which the increments of globally averaged temperature can take place that is the increase of 1°C, 2°C, 3°C and so forth relating it to a wide range of impacts.

A number of these impacts which have been projected tend to pose serious risks to the human societies and things that the people care about, which include water resources, infrastructures, coastlines, human health, food security and the ecosystem of land and ocean.

8.4.1. How Do Scientists Project Future Climate Change?

While determining the future global warming, the biggest factors are projecting the future emissions of Carbon dioxide that is CO_2 and some other greenhouse gases. Furthermore, this depend on how people will produce and can use the energy, what are the national as well as international policies which may be implemented for controlling the emissions and what new technologies might become available.

Scientists try to consider these uncertainties which are done by developing a number of different scenarios of the way in which the future emissions, and also the climate forcing, will evolve.

Each and every scenario from these is based upon the estimations of how different socioeconomic, technological as well as policy factors will changes over the time, which include population growth, economic activity, energy-conservation practices, energy technologies and land use.

The climate models have been used by the scientists to show how the climate system will respond to a number of different scenarios of future greenhouse gas concentration. A number of various models are used and each one of them is developed by a different modeling team.

Each model tends to use a slightly different set of mathematical equations so as to showcase how the atmosphere, oceans, and some other parts of the climate system interact with each other and evolves with time.

The models are compared with one another routinely and also, they are tested against the observations to evaluate the accuracy as well as robustness of model predictions. The most comprehensive suite of modeling experiments to project global climate changes was completed in the year 2005. 23 different models were also included in that from the groups around the world, each of which used the same set of greenhouse gas emissions scenarios.

For all the three future emission scenarios, the projections showed continuous warming, but there were sharp differences in the global average temperature which are evident by the end of the century, having a total temperature increase in the year 2100, in relation with the late 20th century.

This ranges from less than 2°F which is equal to 1.1°C for the low emissions scenario to more than 11°F that is equal to 6.1°C for the high emissions scenario. Such results depict that the magnitude of the future climate change can be largely influenced by the human decisions.

8.4.2. How Will Temperatures Be Affected?

There can be variation in the local temperatures widely from day to day, from week to week and also from season to season, and they so effect on average. The climate modelers have begun to assess the amount of rise in average temperature which might be expected in a number of different regions.

Warming is considered to be greatest in the high latitudes of the Northern Hemisphere and therefore, is considerably large over land in comparison to over ocean. Since, the average temperatures tend to continuously increase, the number of days having a heat index above 100°F is also there to increase throughout this century (the heat index combines the temperature as well as humidity to determine how hot it feels) (Figure 8.3).

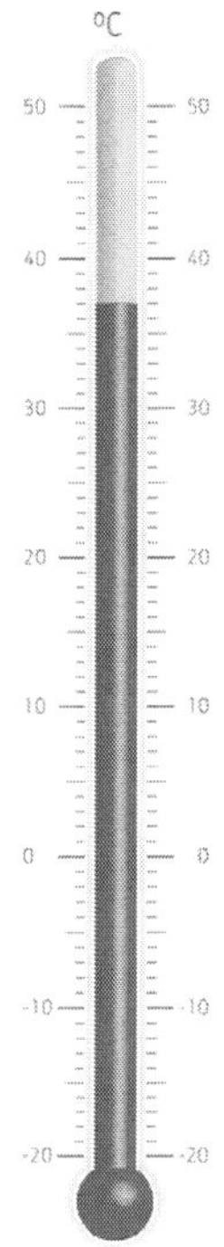

Figure 8.3. Thermometer: Instrument to measure temperature.

Source: Image by Pixabay.

At the end of the century, the center of the United States is expected to experience 60 to 90 additional days per year. In this, the heat index is more than 100°F. Also, the heat waves are expected to last longer along with the increase in the average global temperature. It states that along with the increasing global temperatures, there will also be increase in the risk of the illness and deaths because of heat.

In the similar manner, there is considerable confidence that there will be decrease in the cold extremes, and so there will be decrease in the deaths related to cold.

8.4.3. How Is Precipitation Expected to Change?

Global warming can further intensify the regional contrasts in precipitation which is already present. The dry areas are expected to get even drier, and wet areas are expected to get even wetter. The reason will be the warmer temperatures which tend to increase the evaporation from oceans, lakes, plants, and soil. According to both theory and observations, this increase will boost the amount of water vapor in the atmosphere by about 7% per 1°C (that is equal to 1.8°F) of warming.

Despite the fact that enhanced evaporation provides more atmospheric moisture for rain as well as for snow in some downwind areas, it also dries out the land surface. This exacerbates the effects of drought in some of the regions.

By using the same general approach, which was used for the temperatures, scientists can project regional as well as seasonal percentage change in precipitation which is expected for each 1°C (that is equal to 1.8°F) of global warming.

In the continental United States, the overall pattern of change is much complicated, because it lies in between the drying subtropics of Mexico and the Caribbean and the moistening subpolar regions of Canada. A number of models suggest that there is increase in the drying in the southwestern region of the United States. In many parts of the world, observations show a statistically considerable increase in the intensity of heavy rainstorms.

As per the indications provided by the computer models, along with the warming up of the Earth, this trend will also continue, even in the subtropical regions where there will be decrease in the overall precipitation.

The changes in precipitation will affect the annual stream flow that is approximately equal to the amount of runoff, the water from snow or rain which flows into the rivers and creeks. The global climate models indicate that the runoff that will happen in future is likely to decrease throughout most of the United States, the parts like the Northwest and Northeast are the exceptions, having particularly sharp drops in the Southwest region.

It is expected that there will be decrease in the runoff in some river basins of 5–10% per degree, which include the Arkansas and the Rio Grande. This kind of decrease would be mainly because of the increased evaporation because of the higher temperatures and this will not be offset by changes in the precipitation.

In global terms, stream flow in many temperate river basins, present outside Eurasia, is likely to reduce, especially considering the arid as well as semiarid regions.

It is expected that the increasing temperatures and increasing evaporation and drought can boost the risk of fire in some of the regions. Generally, the forests are already fire-prone, such as the evergreen forests of the Western the United States and Canada.

Such forests are likely to become even more vulnerable to fire along with the increase in temperatures. It is expected that there will be increase in the average area burned by the wildfire per year in the parts of western the United States annually by two or four times per degree of warming.

Simultaneously, there are areas which are dominated by shrubs and grasses, like the parts of the Southwest. These areas may experience a decrease in the fire over time as the temperatures warm up which can cause shrubs and grasses to die out. In such case, the potential societal benefits of fewer fires will be countered by the loss of ecosystems that exists.

8.4.4. How Will Sea Ice and Snow Be Affected?

As there is continuation in the global warming, the many forms of ice on the planet are decreasing in extent, thickness, and duration. There are models which indicate that the ice-free conditions seasonally in the Arctic Ocean are likely to occur before the end of this century. Furthermore, it suggests that in September there is 25% loss in sea-ice extent for each 1°C (that is equal to 1.8°F) in the global warming.

In opposite to the Artic, the sea ice which is present around the Antarctica has expanded on average during the many decades in the past. This increase can be connected with the stratospheric 'ozone hole' over the Antarctic. This further developed due to the use of ozone depleting chemicals in the refrigerants and spray cans (Figure 8.4).

Figure 8.4. Arctic sea ice.

Source: Image by Flickr.

More UV light is allowed from the ozone hole to the lower atmosphere which can be damaging in the Antarctic. This may also lead to lower temperatures as much as heat escapes to space. Nevertheless, it is expected that this effect wane as ozone returns to its normal levels by the time later in this century, due in part to the success of the Montreal Protocol which is an international treaty which have banned the use of ozone-depleting chemicals.

Yet, Antarctic sea ice can further decrease with fewer paces in comparison to the Arctic ice. This is because of the fact that the Southern Ocean stores the heat at greater depths in comparison to the Arctic Ocean, where the heat cannot melt ice much easily. In a number of areas of the world, snow cover may decrease, which snowpack building later in the cold season and melting in the time that is early spring.

When the sensitivity analysis is done, then each 1°C (which is equal to 1.85°F) of the local warming which can result in an average 20% reduction in local snowpack in the western region of the United States. Snowpack has its important implications and that is for drinking water supply and hydropower production.

There are places such as Siberia, some parts of Grassland and Antarctica where the temperatures are low enough so as to provide support to the snow over long periods. In such places, the amount of snowfall can increase even with the shortening of the season. The reason is that the increased amount of water vapor which is connected with the warmer temperatures can enhance the snowfall.

8.4.5. How Will Coastlines Be Affected?

There are some most densely populated regions on Earth which lie at low elevation. This makes the sea level rise and this is a cause for concern. The projection of rise in the sea level continue for centuries in response to human which have caused increase in the greenhouse gases with an estimated 0.5–1.0 meter (that is 20–39 inches) of mean sea-level rise by the year 2100.

However, there is evidence that the increase in the sea-level can be greater in comparison to the expected and this is because of the melting of sea ice. There have a number of studies done recently which have shown more rapid melting from glaciers and ice sheets in comparison to the expected.

Near the top of the range of projections, it is observed that there is rise in the sea-level which were made in the year 1990. Quantifying the threats of upcoming future is challenging and these threats posed to particular coastlines by rising the seas and floods.

There are a number of non-climatic factors which are involved. Some of them can be the place where the people choose to build homes and the risks which will vary much from one location to the next.

Along with this, there is infrastructure damage which is triggered must of the time by extreme events. Some examples of the extreme events can include hurricanes and earthquakes instead of gradual change. Nevertheless, there are some clear hot spots, which are specifically found in large urban areas on the coastal deltas which include those of the Mississippi, Gages, Nile, and Mekong rivers.

In the towns which are along the coast of Alaska reductions in sea ice and melting permafrost is happening which allow the waves to better and erode the shoreline. This is why the relocations are already occurring. Coastal erosion effects at 1.0 meter of sea-level rise which would be much greater and further threatening a number of parts of the US coastline.

8.4.6. How Will Ecosystems Be Affected?

All organism, be it marine be it terrestrial, attempt to acclimate to an environment which is continuously changing or else they tend to move to a more favorable location. But climate change threatens to push some species beyond their ability to either adapt or move. Special focus is being placed on cold-adapted species on the tops of the mountain and at high latitudes.

There are shifts in the timing of seasons and life-cycle events such as blooming, breeding, and hatching and these are causing the mismatches in between species which disrupt the patterns of feeding, pollination, and some other key aspects of food webs. Also, the ability of the species to adapt and move are hampered because of the human infrastructural barriers, competition or interaction with other species and the land use.

Considering the ocean, it can be said that the circulation changes will be the key driver of the ecosystems impacts. According to the satellite data, it shows that the warm surface waters are mixing less with the cooler and deeper waters and separates the near-marine life from the nutrients below. This eventually, reduces the amount of phytoplankton which tends to form the base of the ocean food web.

This problem in the tropics and subtropics is exacerbated by the climate change. However, vertical mixing of water can increase in the temperate and polar waters, along with the expected losses in the sea ice. Simultaneously, the ocean warming will continue to push the ranges of a number of different marine species toward the poles.

The ocean chemistry is changing and this can lead to some other impacts such as warmer waters can result in a decrease in subsurface oxygen, furthermore boosting the risk of the dead zones. The zones where the species high on food chain are majorly not present and this is because of the lack of oxygen. Ocean acidification is when the oceans take in more of the excess CO_2 that is carbon dioxide and this ocean acidification will threaten a number of different species over time, especially coral reefs and mollusks.

But there are some other life forms which will not suffer which include some types of phytoplankton and other photosynthetic organism. These species can benefit from the increase in carbon dioxide. If the emission of carbon dioxide continues in the decades ahead, then this will worsen the ocean acidification.

8.4.7. How Will Agriculture and Food Production Be Affected?

The climate change can have stress on the farming which can threaten the global food security. Despite the fact that there is an increase in the amount of carbon dioxide in the atmosphere which can favor the growth of a number of different plants, it is not required to translate it into more food. In the higher temperatures, crops tend to grow more quickly which result in shorter growing period and less time to produce grains (Figure 8.5).

Figure 8.5. Agriculture.

Source: Image by Wikimedia Commons.

Along with this, the change in climate will bring other hazards which include greater water stress and also, the risk of higher temperatures peaks which in turn can damage the crops quickly. Across the regions and by crop, there can be variation seen in the agricultural impacts.

Crop and pasture lands in the middle to high latitudes can get benefitted because of the moderate warming, relative increase in carbon dioxide and also changes in precipitation. But there will be decrease in the yield in the areas which are seasonally dry and at low latitude.

Let's take an example California. In California, half of the fruit and vegetable crops of the nation are grown. Here, the climate change is projected so as to decrease the yields of almonds, avocados, walnuts, and table grapes by up to 40% by the year 2050.

Regional assessment was also done for other parts of the world which consistently conclude that the climate change can cause serious risks to the critical staple crops in the regions of sub-Saharan African and in the places, which tend to rely on the water resources from glacial melt and snowpack.

As per the indications provided by the modeling, the benefits related to CO_2 for some crops will majorly be outweighed by some of the negative factors if in case, there is increase in the global temperature more than 1°C (that is equal to 1.8°F) from the late 20th century. There can be some impacts which are mentioned below:

- For each degree of warming, in the United States and Africa, yields of corn and in India, yields of wheat is dropped by 5–15%;
- Crop pests, weeds, and disease shift in geographic range and frequency, and;
- If 5°C (that is equal to 9°F) of global warming were to be reached, then yield losses will be experienced by most regions of the world and also, the global grain prices would potentially double.

In the prosperous areas, growers may be able to adapt to these threats. Let's take an example of such threats that is by varying the crops which is grown by them and the times at which they are grown. However, adaptation may be not much effective where local warming exceeds by 2°C (that is equal to 3.6°F) and will be limited in the tropics, where the growing season is restricted by moisture and not temperature.

8.5. CLIMATE CHANGE AND AGRICULTURE

The climate change can impact the agriculture and this will be one of the major deciding factors which influence the future food security of mankind on earth. Agriculture is sensitive to climate change and also, it is one of the major drivers for climate change.

Over a time, understanding the weather changes and adjusting the management practices towards achieving better harvest are the major challenges to the growth of agricultural sector as a whole.

The sensitivity of agriculture towards climate is not certain, because there is regional variation in rainfall, temperature, crops, and cropping systems, soils, and management practices. The inter-annual variations in the temperature as well as the precipitation were much higher in comparison to the predicted changes in the temperature and also, in the precipitation.

Every crop is different and every crop responds differently as the global warming will have a complex impact. The tropics are said to be more dependent on the agriculture because of the fact that 75% of the population of the world lives in tropics and around two third of these people have agriculture as their main occupation.

Any impact on the tropical agriculture will affect their livelihood because of the fact that they have low levels of technology, wide range of pests, diseases, and weeds, land degradation, rapid population growth and unequal distribution of land.

There are six major crops in the world such as wheat, rice, sorghum, maize, barley, and soybean which are grown in 40% cropped area and contribute around 55% of non-meat calories and above 70% of animal feed (FAO, 2006). If there is any effect on these crops then it would definitely affect the food security adversely.

Main projections for climate change at Global Level: The projections of future climate patterns are mainly based on the models which are computer-based that is related to the climate system. These models incorporate the important factors and process of the atmosphere as well as the oceans, which include the expected growth in greenhouse gases considering the socio-economic scenarios for the upcoming decades.

There were published results from a number of different models which was examined by the IPCC and on the basis of evidence; IPCC has estimated that by the year 2100:

- The global average surface warming (that is surface air temperature change) will increase by 1.1–6.4°C.
- The oceans will become more acidic.
- The sea level will rise between 18 cm and 59 cm.
- It is much likely that at higher latitudes, there will be more precipitation and it is likely that in most subtropical land areas, there will be less precipitation.
- It is likely that tropical cyclones (such as typhoons and hurricanes) will become more intense, with larger peak wind speeds and heavier precipitation related with ongoing increases of tropical sea surface temperatures.
- It is much likely that hot extremes, heat waves and heavy precipitation events will continue to become more frequent.

> Global food security depends on both sufficient food production as well as food access, and so, is defined as a state when:
> 'all people, at all times, have physical as well as economic access to sufficient, safe, and nutritious food to meet their dietary needs and food preferences for living an active and healthy life' (FAO, 1996).

8.6. IMPACT OF CLIMATE CHANGE ON AGRICULTURE

There is a principal barrier to food security and that currently is food access. Globally, the sufficient food is produced so as to feed the current world population, and yet there are more than 10% of the populations who are undernourished.

It is much likely that the climate change will contribute substantially in the food insecurity in the upcoming future and this can be done by increasing the food prices and decreasing the food production. Food can become much more expensive along with the increase in the energy prices due to the climate change mitigation efforts.

Water required for the production of food can become scarcer because of the increase in crop water use and drought. There can be increase in the competition for land because certain areas become climatically not suitable for production.

Along with this, there are extreme weather events, in relation with the climate change which can cause sudden reductions in agricultural productivity and this has resulted in rapid increase in the prices. Let's take an example. In the summer of the year 2010, the heat waves resulted in yield losses in key production areas which include Ukraine, Russia, and Kazakhstan and contributes to a dramatic increase in the prices of the staple foods.

These rising prices emphasized on growing numbers of local people into poverty, and further providing a sobering representation of how the food insecurity is influenced by the climate change.

As per the consensus of the Intergovernmental Panel for Climate Change (IPCC), since the period of 1950s, the substantial climate change has already occurred and also, it is likely that the global mean surface air temperature will increase by 0.4–2.6°C in the second half of this century (this depends on the future greenhouse gas emissions).

The major source of greenhouse gas emissions is agriculture and the wider food production system. There can be future intensification of agriculture to compensate for the decreased production (which is caused by climate change) along with increasing demand for animal products.

This can further increase such emissions. As per the estimation, it can be said that there is demand for livestock products which will grow by +70% between the year 2005 and the year 2050.

The gradual increase in the temperature and carbon dioxide can result in much favorable conditions which could eventually increase the yields of some crops. There are some regions, where these potential yield increases are much likely to be restricted by some of the extreme events specifically extreme heat and drought during the crop flowering time. During the 21st century, crop production is expected to decrease in a number of areas because of the climate changes.

Heat waves are the periods of extreme high temperature. These heat waves are much likely to become more frequent in the near future and it represents a major challenge for the agriculture. Heat stress can be caused due to heat waves in both animals as well as plants and can also, have negative impact on the food production (Figure 8.6).

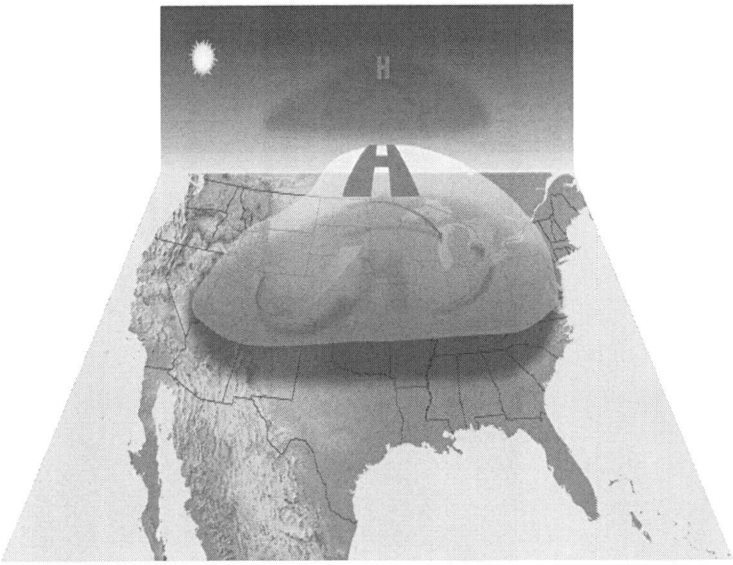

Figure 8.6. Heat wave.

Source: Image by Wikimedia Commons.

Specifically, extreme periods of the high temperature are harmful for the crop production, if they occur at the time of crop flowering. If in case, this single stage is disrupted, then the critical stage will also get disrupted and there can be times when there may be no seeds at all.

Considering the animals, it can be said that the heat stress can result in low productivity as well as low fertility. It can also have its negative effects on the immune system as it can make them more prone to certain diseases.

There is evidence for an increase in the heat waves and these exists from warming which have already been occurred and greater than the increase that was expected in the heat wave frequency and magnitude. It is much difficult to make accurate predictions frequency as well as magnitude of heat waves. But there exists a consensus among the projections that measurements for both will continue to increase in the UK, in Europe and also at a global scale.

The heat waves leave an impact which is expected to be non-uniform, having disproportionately negative effects in not so developed countries. In addition to the other aspects of climate change which include increased drought incidence, they can exacerbate existing issues around the food security.

The changes which are projected in the climate are not limited to increases in temperature and heat waves and also, the large changes in rainfall patterns are expected to occur.

There are some regions which are likely to suffer from more droughts in the upcoming future, but there are some other regions are expected to face the opposing issues of torrential rains and increased flooding. Considering the coastal areas, there can be rise in the sea levels which can result in complete loss of agricultural land.

More problems can be the result of the warmer climates from pests and diseases, and shifts in the geographical distribution of some specific pests. Let's take an example. There are many insects which serve as a vector for disease transmission and these insects are likely to migrate further pole-ward in the upcoming future, where livestock have so far not been exposed to such diseases.

There are experimentations done on a number of crops which helps in defining the responses of yield to various stresses. It is important to quantify these responses and identify when the agriculture is most vulnerable to stress and this proves to be beneficial in helping to identify the most efficient strategies for adaptation.

It is expected that the crop-level adaptation to climate change is the key in minimizing the future yield losses and can include: sowing time, changing crop cultivars, irrigation practices, and cultivation techniques.

There are researches going on which addresses the challenges of maintaining and increasing the crop production under global change. From climate change and extreme weather events, there can be some risks which have been identified and the strategies have suggested helping in maintaining the production.

Restoring farm type, crop, or cultivar scale diversity into food systems have been included in them, to improve their resilience as well as making crop improvements which in turn can enhance stress tolerance. There can be some other strategies which can include developing pre-defined, international responses to food shortages for further preventing the food price shocks which might reduce the people's access to food.

8.7. HOW WE CAN USE AGRICULTURE TO FIGHT CLIMATE CHANGE

In the history of this world, the year 2019 will go down and will be considered as the most difficult planting season for farmers of North American, with more than 10 million acres of crops going unplanted because of the extreme weather conditions.

Simultaneously, in India, the farmers in Punjab are experiencing the rain showers almost every month and air that is more humid is resulting in the greater pest infestations for the first time in its history.

The effects of the climate change can be felt on a daily basis, especially by farmers, but there are very few solutions which have been discussed to address this catastrophic threat. Nevertheless, there is one solution which is widely known to reduce the amount of greenhouse gasses which are trapped in the atmosphere and that are agriculture.

The agriculture's footprint is reduced and also, excess carbon is captured which is generated by some other industries and this happens by reducing tillage, planting cover crops, expanding crop rotations and reintegrating livestock into crop production systems.

The carbon that is captures is further converted into plant material and soil organic matter. This in turn improves the soil health and increases the ability to produce the food on land in the upcoming future. These practices are also able to reduce the input costs.

A greener production system is pursued by the farmers. This requires the farmers to embark on uncharted territories having no guarantee of immediate success. Usually, the farmers tend to experience decrease in the yields during the transition process. This is because they gain the required experience to learn and perfect the implementation of more regenerative as well as beneficial practices.

The decrease in production tends to pose a financial issue which is difficult to overcome, especially considering the Indian farmers, who already face a hard time while competing with the developed nations, as the subsides have artificially driven down the prices of agriculture produce.

The import and export policy decision taken by the government have heavily favored the consumers over producers by keeping the prices artificially low. They also have a large impact on the ability of farmers living in India so as to adopt more sustainable practices.

Let us consider the case of US; the farmers tend to face systematic financial challenges such as facing difficulty in assessing the sustainable inputs at some reasonable price. The farm input suppliers are highly concentrated, which further exert the considerable pricing power and making systems innovation unattractive to their bottom line. The results formed as high operating costs along with the upfront costs which is required increases the need for access to external capital.

However, the capital is considered to be mostly available to farmers with the production systems which are most traditional as well as low-return. In other words, it can be said that, generally the systems which are least regenerative, tend to emit an abundance of greenhouse gases. This can result in the most land degradation, and so, are the most likely to have access to capital.

Subsidies as well as regulation play a critical role in the availability of external capital. The reason is that these cash streams can serves as a de-risking mechanism for finance players. The policies of the US government include federally subsidized crop insurance.

Such policies are focused on ensuring a stable food supply instead of the nutritional value or environmental impacts of the food that is being produced.

This emphasizes on quantity over quality which is the same for the case of India. Here, the decision makers have rolled out an initiative so as to increase the food production by the year 2050. This aim of this initiative has become the cornerstone of national policy of India and the metric for measuring success of farmers.

There is one more key challenge that is faced is the lack of data and available information on these management practices and also, the lack of ways to measure which of these new, innovative systems which in turn store most of the carbon in the soil of Earth. Such kind of tool can open the door to private parties' further compensating farmers to sequester carbon.

In the meantime, the regulations of government and also, growers' low profit margins, stifle farmer innovation, further limiting the ability for new as well as creative thinkers to join the industry or for current farmers for testing new practices.

Let's take an example. In the US, the subsidized crop insurance inflates the value of farmland as well as locks producers into a low risk, low reward system. This makes it hard for the new smallholder farmers to enter the business or current farmers to walk away from the easy revenue.

In addition to this, in the US, the selling or exchanging farms and farmlands have become financially efficient only after death which is made by the tax code. Along with this, the farmland investments can be further used for providing shelter to the gains realized on non-agriculture real estate, along with the increasing barrier to entry.

In both the US as well as India, there is a lack of financial incentives that is for farmers to pursue innovation, whereas, the public funding for the fundamental research as well as the application of research to agricultural practices have been decreased in the real terms.

The public research system has starved of funds and this has taken the easier path. These systems have focused on identifying the ways for maximizing the farm yields by using the mono-cropping and using the chemicals. Both of this tends to increase the agriculture emissions.

There is one way to incentivize the farmers so that they focus on the increasing soil health. That way is through de-commoditizing production and further, making it easier to identify as well as track the producers because the food makes its way from farm to table.

By doing this, consumer demand can be unlocked as an incentive for greener farming. If consumers are willing to pay a higher price for the food products which have been produced in an environmentally friendly manner, the economics of the system can be turned on its head.

Unfortunately, the crop buyers in US currently blend and commoditize the production, further leaving no pathway for consumers so as to give reward farmers to establish more favorable production systems.

In India, the disposable income is fairly limited. The price of foods increases which could limit the ability of individuals so as to purchase foods, the fear of food inflation is amplified, a problem that the government dreads.

Each and every country, climate, and plot of land poses their challenges and this is particularly clear that there is no 'one size fits all' solution. Nevertheless, it is important to address these barriers to unlock the ability of agriculture so as to solve a portion of the climate change problem.

8.8. CONCLUSION

When the climate change occurs, then there is dramatic increase in the temperatures. When there is increase in the temperature, then there are a number of changes that takes place on this Earth.

The conclusion that can be drawn from the chapter is that it is important to take part and try to stop global warming and some other effects on climate change. If the temperature of earth continues to rise in the upcoming future, then the living things on earth would become extinct because of the high temperatures.

For controlling the global warming, if the contribution of the humans is made then, this world would be cooler as well as the high temperatures will decrease that is currently there. If everybody tries to end most of the climate changes which are occurring, then this world would be a safer place to live on.

REFERENCES

1. Climate Change and its Impact on Agriculture, (n.d.). [eBook] p. 2. Available at: https://www.manage.gov.in/studymaterial/CCA-E.pdf (accessed on 2 April 2020).
2. Dittmann, M., (n.d.). Impact of climate change on agriculture. [online] FutureLearn. Available at: https://www.futurelearn.com/courses/climate-smart-agriculture/0/steps/26565 (accessed on 2 April 2020).
3. Impacts, Vulnerabilities and Adaptation in Developing Countries, (2007). [eBook] p. 1. Available at: https://unfccc.int/resource/docs/publications/impacts.pdf (accessed on 2 April 2020).
4. Riensche, B., & Vir, J. A., (2019). Here's How We Can Use Agriculture to Fight Climate Change. [online] World Economic Forum. Available at: https://www.weforum.org/agenda/2019/09/here-s-how-we-can-use-agriculture-to-fight-climate-change/(accessed on 2 April 2020).
5. UKEssays.com. (2018). Climate Change Essay. [online] Available at: https://www.ukessays.com/essays/environmental-sciences/causes-of-climate-change.php (accessed on 2 April 2020).
6. Warming, Climate Changes, and Impacts in the 21st Century and Beyond, (n.d.). [eBook] p. 4. Available at: http://nas-sites.org/americasclimatechoices/files/2012/06/19014_cvtx_R1.pdf (accessed on 2 April 2020).

INDEX

F

G

H

I